Functional Safety in the Process Industry

A Handbook of Practical Guidance in the Application of IEC61511 and ANSI/ISA-84

KJ Kirkcaldy, D Chauhan

First published 2012.

ISBN 978-1-291-18723-6

CONTENTS

ABBREVIATIONS

λ	Failure rate, the ratio of the total number of failures occurring in a given period of time
λD	failure rate of dangerous failures
λDD	failure rate of dangerous failures detected by diagnostics
λDU	failure rate of dangerous failures undetected by diagnostics
λS	failure rate of safe failures
1oo1	1 out of 1 voting (Simplex)
1oo2	1 out of 2
AI	Analogue Input
ALARP	As Low As Reasonably Practicable
ANSI	American National Standards Institute
BMS	Burner Management System
BPCS	Basic Process Control System
C&E	Cause and Effect
CBA	Cost Benefit Analysis
CCF	Common Cause Failure
COMAH	Control Of Major Accident Hazards
Dangerous failure	This is a failure mode that has the potential to put the safety-related system into a hazardous or fail-to-function state
DCS	Distributed Control System
DD	Dangerous Detected
DI	Digital Input
DO	Digital Output
DU	Dangerous Undetected
E/E/PES	Electrical / Electronic / Programmable Electronic System
ESD	Emergency Shutdown
ESDV	Emergency Shutdown Valve
F&G	Fire and Gas
f/hr	Failures per hour
FC	Fail Closed
FDS	Functional Design Specification
FMECA	Failure Modes, Effects and Criticality Analysis
FO	Fail Open
FPL	Fixed Programmable Language

ABBREVIATIONS

FSC	Functional Safety Capability
FVL	Full Variability Language
HAZAN	Hazard Analysis
HASAW	Health and Safety at Work Act (HSW)
HAZOP	Hazard and Operability Study
HFT	Hardware Fault Tolerance
HIPPS	High Integrity Pressure Protection System
HSE	Health and Safety Executive
I/O	Input/Output
IEC	International Electrotechnical Commission
IPL	Independent Protection Layer
ISA	International Society of Automation
LOPA	Layer of Protection Analysis
LVL	Limited Variability Language
MDT	Mean Down Time
MooN	M out of N (general case)
MTBF	Mean Time Between Failures
MTR	Maximum Tolerable Risk
MTTF	Mean Time To Failure
MTTR	Mean Time To Repair
Non-SR	Non-Safety Related
O&M	Operation and Maintenance
OPSI	Office of Public Sector Information
P&ID	Piping and Instrumentation Diagram
PA	Per Annum
PE	Programmable Electronics
PFD	Probability of Failure on Demand
PFH	Probability of Failure per Hour
PSD	Process Shutdown
PT	Pressure Transmitter
PTI	Proof Test Interval
QMS	Quality Management System
R2P2	Reducing Risk Protecting People
RBD	Reliability Block Diagram
RRF	Risk Reduction Factor

ABBREVIATIONS

S	Safe
SA	Safety Authority
Safe failure	This is a failure mode which does not have the potential to put the safety-related system in a hazardous or fail-to-function state.
SFF	Safe Failure Fraction.
SIF	Safety Instrumented Function
SIL	Safety Integrity Level.
SIS	Safety Instrumented System
SOV	Solenoid Operated Valve
SRS	Safety Requirements Specification
STR	Spurious Trip Rate
TMR	Triple Modular Redundant
Tp	Proof Test Interval

Forward

The safety standard IEC61508 covers the safety management of electrical, electronic and programmable electronic systems throughout their lives, from concept to decommissioning. It brings safety principles to the management of systems, and safety engineering to their development.

At its core is the principle that in safety planning, safety goals based on risk assessment should be set, and then that the rigour of management and processes should be appropriate to meeting them. This makes the standard goal-based rather than prescriptive, and means that compliance with the standard does not exonerate users of any blame in the event of a safety problem.

IEC61508 is intended both as the basis for the preparation of more specific standards and also for stand-alone use. However, the former application is preferred; the latter will require tailoring, significant understanding of it by management, and considerable planning in its introduction and use.

For many, the standard is difficult to read and understand but it has already been hugely influential and will continue to be the basis of modern safety standards and legal frameworks. It is essential therefore, that all with responsibilities at any stage of the life of a safety-related system should make the effort to understand it.

The safety standard IEC61511 is the process industry specific implementation of IEC61508 and this book aims to provide an introduction to, and guidance in its application. The American Standard ANSI/ISA-84.00.01 is based on IEC61511 and is essentially identical to it. This guidance therefore applies to both.

The purpose of this book is to enable a better understanding of the standards and their requirements to be gained. The book aims to use simple language, illustrated with worked examples from actual projects, to explain the basic principles and requirements together with techniques that could be used to meet those requirements.

Disclaimer

Whilst the techniques presented here have been used successfully in demonstrating compliance on actual projects, it should be noted that compliance, the techniques used to demonstrate compliance and the collation of supporting evidence remains the responsibility of the duty holder.

The use of the square bracket [   ] indicates a cross-reference to a section within this document.

1. INTRODUCTION

Background

IEC61508 [20.1] is an international standard published by the International Electrotechnical Commission (IEC) and its primary objective is to address aspects to be considered when electrical, electronic or programmable electronic (E/E/PE) systems are used to perform safety functions.

IEC61508 is a generic standard that applies to all E/E/PE safety-related systems, irrespective of their use or application. The title of the standard is:

> **IEC61508:2010 Functional Safety of Electrical / Electronic / Programmable Electronic Safety-Related Systems.**

The standard is founded upon the primary principle that there is a process that may pose a risk to safety or the environment, should something go wrong with the process or equipment. The standard is consequently aimed at process upsets and system failures, as distinct from health and safety hazards such as trips and falls, and allows process safety to be managed in a systematic, risk-based manner.

The standard assumes that safety functions are to be provided to reduce those risks. Safety functions may together, form a Safety Instrumented System (SIS) and their design and operation must be based on an assessment and understanding of the risks posed.

A secondary objective of IEC61508 is to enable the development of E/E/PE safety-related systems where application sector standards may not exist. Such second tier guidance in the process industry is covered by international standard IEC61511 [20.2]. The title of this standard is:

> **IEC61511:2004 Functional Safety – Safety Instrumented Systems for the Process Industry Sector.**

IEC61511 is not a design standard but a standard for the management of safety throughout the entire life of a system, from conception to decommissioning. Fundamental to this approach is the overall safety lifecycle which describes the activities that relate to the specification, development, operation or maintenance of a SIS.

What is Functional Safety?

IEC61511-1, 3.2.25 provides the following definition.

"Functional Safety is the part of the overall safety relating to the process and the Basic Process Control System (BPCS) which depends on the correct functioning of the SIS and other protection layers."

More simply, functional safety is the risk reduction provided by the functions implemented to ensure the safe operation of the process.

The Structure of IEC61511 (the Standard)

The standard, IEC61511 consists of three parts.

Part 1 outlines the requirements for compliance. Project planning, management, documentation, and requirements for competence, as well as the technical requirements for achieving safety throughout the safety lifecycle are defined.

In general, Part 1 is 'normative' in that it defines specific requirements for compliance and is laid out in a consistent structure to allow a clause by clause demonstration of compliance.

Part 2 provides guidance on the use of Part 1.

Part 3 gives worked examples of risk assessment leading to the allocation of safety integrity levels, [4].

Parts 2 and 3 are 'informative' and provide guidance on the normative requirements.

The structure of the standard is illustrated by Figure 1.

FIGURE 1. STRUCTURE OF THE STANDARD

Technical Requirements	Support Parts

Part 1
IEC61511-1, 8
Development of the overall safety requirements (concept, scope definition, hazard and risk assessment)

↓

Part 1
IEC61511-1, 9, 10
Allocation of the safety requirements to the safety instrumented functions and development of safety requirements specification

↓

Part 1
IEC61511-1, 11, 12

Design phase for safety instrumented systems	→ ←	Design phase for safety instrumented system software

↓

Part 1
IEC61511-1, 13, 14, 15
Factory acceptance testing, installation and commissioning and safety validation of safety instrumented systems

↓

Part 1
IEC61511-1, 16, 17, 18
Operation and maintenance, modification and retrofit, decommissioning or disposal of safety instrumented systems

Part 1
IEC61511-1, 2: References
IEC61511-1, 3: Definitions and Abbreviations
IEC61511-1, 4: Conformance
IEC61511-1, 5: Management of Functional Safety
IEC61511-1, 6: Safety Life-Cycle Requirements
IEC61511-1, 7: Verification
IEC61511-1, 19: Information Requirements
IEC61511-1, Annex A: Differences

Part 2
IEC61511-2
Guideline for the Application of Part 1

Part 3
IEC61511-3
Guidance for the Determination of the Required Safety Integrity Levels

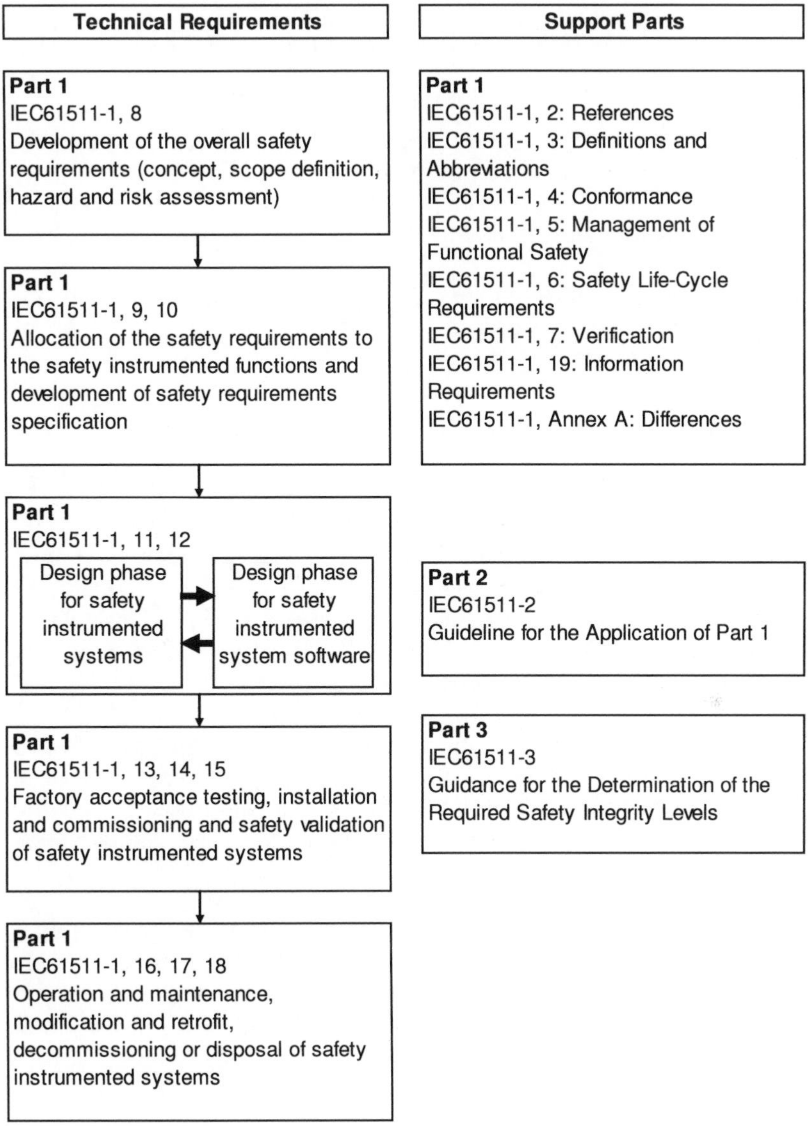

The International Electrotechnical Commission (IEC)

The International Electrotechnical Commission was founded in 1906, with British scientist Lord Kelvin as its first president, and is based in Geneva, Switzerland. It prepares and publishes International Standards for electrotechnology, i.e. electrical, electronic and related technologies.

The IEC supports the safety and environmental performance of electrotechnology, promotes energy efficiency and renewable energy sources, and manages the conformity assessment of equipment, systems or components to its International Standards.

The standard and all other IEC publications are protected and are subject to certain conditions of copyright but can be purchased or downloaded from the IEC.

Compliance with IEC61511

Requirements of the Health and Safety at Work Act etc. 1974

The Health and Safety at Work Act etc. 1974 (HASAW, or HSW) [20.3] is the primary legislation covering occupational health and safety in the UK. The Health and Safety Executive (HSE) is responsible for enforcing the Act and other Acts and Statutory Instruments relevant to the working environment.

The full text of the Act can be obtained from the Office of Public Sector Information (OPSI) or downloaded free of charge. Users of legal information must exercise a degree of caution. Printed or on-line documents may not be current and therefore users should seek independent legal advice or consult the HSE Infoline.

Put in simple terms, the Health and Safety at Work Act states that it shall be the duty of every employer to ensure, so far as is reasonably practicable, the health, safety and welfare at work of all his employees. This includes the provision and maintenance of plant and systems of work that are, so far as is reasonably practicable, safe and without risks to health.

In addition, it shall be the duty of every employer to conduct his undertaking in such a way as to ensure, so far as is reasonably practicable, that persons not in his employment who may be affected, are not thereby exposed to risks to their health or safety.

Requirements for Compliance

IEC61511 states that to claim compliance it shall be demonstrated that the requirements of the standard have been satisfied to the required criteria and, for each clause or subclause, all the objectives have been met.

In practice, it is generally difficult to demonstrate full compliance with every clause and subclause of the standard and some judgement is required to determine the degree of rigour which is applied to meeting the requirements. Typically, the degree of rigour required will depend upon a number of factors such as:

- the nature of the hazards;

- the severity of the consequences;

- the risk reduction necessary;

- the life-cycle phase that applies;

- the technology involved;

- the novelty of the design.

In other words, a risk-based decision must be made. Where there is a lack of experience, some external involvement would add to the credibility of the claim.

Consequences of Non-Compliance

The standard is not law and therefore, whether you comply with its requirements or not, you should be aware of the consequences of non-compliance. As an employer, duty holder or risk owner, you have an obligation under the Health and Safety at Work Act to manage risk at your place of work.

The standard does provide a systematic approach to managing all safety lifecycle activities for systems that are used to perform safety functions and therefore is a good source of information and techniques. Should something go wrong that results in someone being hurt or becoming ill and you didn't use the best information available to you in managing that risk, then you would be at risk from investigation and prosecution under the Health and Safety at Work Act.

The information you collate and the analysis that you provide in meeting the requirements of IEC61511, effectively becomes your defence in court should something go wrong.

Requirements for Compliance on New Plant

It is clear, that if you are involved in any part of the safety lifecycle, it would be reasonable to expect you to apply the best information available in ensuring that the risks associated with your plant are managed to a tolerable level. It could be argued that the best information available is IEC61511, and therefore, should something go wrong, failure to comply could be construed as negligence.

Requirements for Compliance on Existing Plant

There are many plants that were designed and built before IEC61511 was formally published and generally available. This situation doesn't change your responsibilities however and if you are involved in any part of the safety lifecycle of an old plant, e.g. operation, maintenance etc., then your obligations under the Health and Safety at Work Act remain, and the risks should still be managed accordingly. The standard therefore, still applies to those old plants.

ANSI/ISA-84 [20.4] specifically addresses legacy systems by stating that for an existing SIS, designed and constructed in accordance with the codes, standards, and practices applicable prior to the issue of the standard, the owner/operator shall determine that the equipment is designed, maintained, inspected, tested, and operating in a safe manner. In effect, you must verify that your existing systems are safe using the best methods available to you.

In reality, you may feel you need to go back to the early parts of the safety lifecycle for the existing plant and revisit or even conduct a new Hazard and Operability (HAZOP) Study from scratch. Taking the process through to its conclusion, you may identify risks that are not protected by existing safety functions and it will be your responsibility to manage those risks in some way.

In all probability, it will not be cost effective to engineer new Safety Instrumented Functions (SIFs) for a 20 year old plant. However, if your plant has been operating safety for a reasonable length of time, then the risks that you identify and their likelihoods, taking into account any existing safeguards, may already be tolerable.

Your obligation is at least to document the process: to make sure all hazards have been identified, the risks assessed and the protective functions or safeguards that currently exist, evaluated for their effectiveness. In this situation, you have the benefit of hindsight and you can quantify your hazard frequencies more accurately, using your own historical records, than you would be able to if it was a new installation. You should therefore be able to demonstrate by analysis that the risks you have identified are tolerable.

If the worst comes to the worst, and you arrive at the situation where there are some unprotected hazards, or some additional risk reduction measure is required, then you need to know this and you should take steps to achieve this.

Reasons for Compliance with IEC61511

Apart from the implied legal obligation under the Health and Safety at Work Act there may be other reasons for complying with the standard:

- Contractual requirements;
- Optimisation of design architecture;
- Possible marketing advantage.

It could be argued that the first duty of a business is to survive and its objective should not be the maximisation of profit, but the avoidance of loss. On that basis, you must ask yourself whether you would rather learn from the mistakes of others, or make them all yourself.

Applying IEC61511

Functional Safety can only be applied to complete functions which generally consist of a sensor, a computer or PLC, and an actuated device. It is meaningless to apply the term to products: items of equipment such as sensors, or computers.

Therefore when a manufacturer claims that their product is a SIL2 Pressure Sensor, or a SIL3 PLC for example, in reality, this means that the Pressure Sensor is *suitable for use* in a SIL2 safety function, or the PLC is *suitable for use* in a SIL3 safety function. The manufacturer should qualify the claims with caveats and restrictions on its use, such as the requirements for fault tolerance or proof testing for example, in order to achieve the claimed SIL.

The claims of the manufacturer may even be backed up with a SIL Certificate issued by an independent assessment body but this does not mean that the out-of-the-box safety function will be SIL compliant. The SIL Certificate is not a substitute for demonstrating compliance and the duty holder cannot use such product claims to discharge his responsibilities under the Health and Safety at Work Act.

Do I have to comply with the standard?

New build

As previously stated, there is an implied legal obligation to comply with the standard. This means that the standard is not law but the law does require the duty holder, or risk owner, to manage risk to an acceptable level. The standard provides a systematic approach to achieve this and therefore should something go wrong that results in someone being hurt, then failure to use the best information available could be viewed as an indication of negligence and may result in prosecution.

Existing Plant

For existing plant, the Health and Safety at Work Act still applies and therefore risks should still be identified and managed appropriately. IEC61511 still provides an applicable model for managing risks on legacy plant designed and operated before the standard was published.

2. OVERALL SAFETY LIFECYCLE

Safety Lifecycle

The Safety Lifecycle encapsulates all the necessary activities from specification, development, operation or maintenance of the SIS. Depending upon your scope of activities, you may only be involved in some of the phases, e.g. operation and maintenance, but you should be aware of the whole lifecycle approach.

The overall safety lifecycle is presented in Figure 2.

FIGURE 2. IEC61511 SAFETY LIFECYCLE

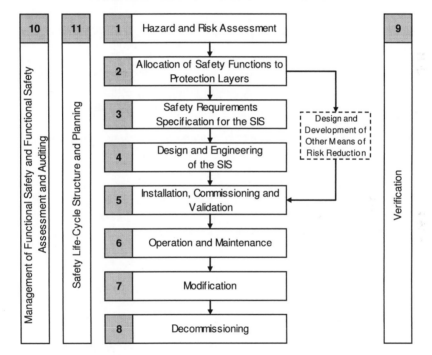

Lifecycle Phases

Each phase of the lifecycle describes an activity, for which you should have documented procedures. Each activity has information requirements as inputs and produces information as outputs for use in subsequent phases.

Lifecycle Phase 1 defines the scope in terms of physical, social and political boundaries and addresses the safety implications in terms of hazards and the perception of risk. This is fundamental to understanding the hazards and risks posed by the process. Once the necessary risk reduction has been determined, the means of achieving it are specified during the allocation, Lifecycle Phase 2 and the Overall Safety Requirements, Lifecycle Phase 3.

In Lifecycle Phase 4, the Overall Safety Requirements are engineered as safety functions. Optimisation of functions, separation and other design issues such as testing philosophy are examined at this point and the planning for these activities is addressed as part of Lifecycle Phase 11.

Lifecycle Phases 5 to 10 demonstrate that the standard is not restricted to the development of systems, but covers functional safety management throughout the life of a system. Many of the requirements in the standard are technical in nature, but the lifecycle approach places equal importance on effective management activities such as planning, documentation, operation, maintenance and modification and these must be included in all phases.

Documentation, management and assessment activities lie in parallel to, and apply to all of the lifecycle phases and activities shown in Figure 2.

It should be noted that although the standard describes lifecycle phases and information requirements for each phase, in practice some of the phases and their associated documents may be combined if appropriate. Clarity and simplicity are important and the activities should be performed and information presented in the most effective manner.

Compliance Requirements

Because the standard is non-prescriptive, compliance is never straightforward. How much or how little you do in claiming compliance is personal choice but you should satisfy yourself that you have done enough. A clause by clause compliance approach is recommended to be sure that you have considered everything that would be reasonably expected of you. In other words, that a rigorous approach has been adopted.

Compliance to the standard requires you to demonstrate, with evidence, that a systematic approach has been adopted to managing risk and that approach has been applied over the appropriate parts of the lifecycle. This systematic approach, provided by the standard, is based on the safety life-cycle.

Compliance with the standard requires that the lifecycle is understood and the required activities are performed and documented. Following the lifecycle is not a paperwork exercise that can be satisfied with the generation of reports, documents and boxes to be ticked. Compliance requires the activities to be carried out in an effective manner and information to be produced at each phase that enables subsequent phases to be carried out.

A limited scope of activity rarely applies and it is recommended that all phases of the lifecycle are considered. For example, for an operator, modification in the operation and maintenance phase, may require earlier decisions and assessments, e.g. risk analysis to be re-assessed by reverting back in the lifecycle.

Safety Lifecycle Phases 1 and 2

Figure 3 shows the activities and information requirements for Lifecycle Phase 1 (Hazard and Risk Assessment) and Lifecycle Phase 2 (Safety Requirements Allocation). The figure shows the information required as an input (I/P) to the activity, and the information produced by the activity as an output (O/P) for use in the subsequent phase.

The output of Phase 1 will generally be a HAZOP and risk analysis, identifying safety function requirements and risk reduction targets.

Phase 2 deals with the allocation of safety functions based on the safety requirements identified in the previous phase. Safety requirements allocation is the process of addressing each of the safety requirements and allocating safety instrumented functions. This is an iterative process and will take into account the process and other risk reduction measures that may be available to meet the overall safety integrity requirements.

FIGURE 3. SAFETY LIFECYCLE PHASES 1 AND 2

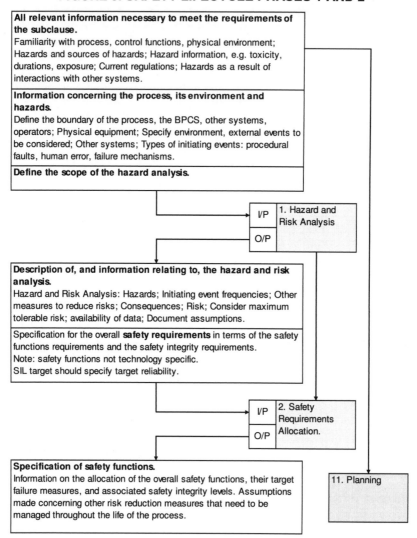

It is important that when the safety function allocation begins, forthcoming phases including installation, commissioning and validation, operation and maintenance are also planned (also refer to Figure 6).

Safety Lifecycle Phases 3 and 4

Lifecycle Phase 3 addresses the Safety Requirements Specification (SRS) which enables the Design and Engineering Phase (Phase 4) to begin, Figure 5.

FIGURE 4. SAFETY LIFECYCLE PHASES 3 AND 4

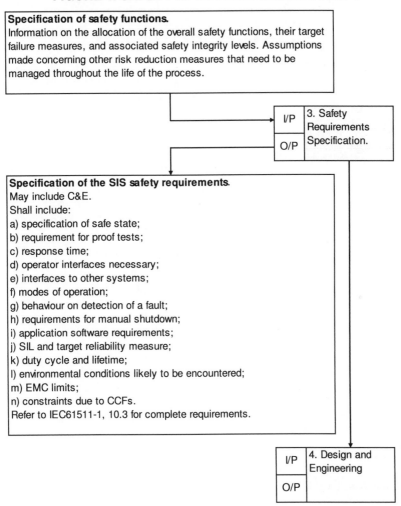

Specification of safety functions.
Information on the allocation of the overall safety functions, their target failure measures, and associated safety integrity levels. Assumptions made concerning other risk reduction measures that need to be managed throughout the life of the process.

| I/P | 3. Safety Requirements |
| O/P | Specification. |

Specification of the SIS safety requirements.
May include C&E.
Shall include:
a) specification of safe state;
b) requirement for proof tests;
c) response time;
d) operator interfaces necessary;
e) interfaces to other systems;
f) modes of operation;
g) behaviour on detection of a fault;
h) requirements for manual shutdown;
i) application software requirements;
j) SIL and target reliability measure;
k) duty cycle and lifetime;
l) environmental conditions likely to be encountered;
m) EMC limits;
n) constraints due to CCFs.
Refer to IEC61511-1, 10.3 for complete requirements.

| I/P | 4. Design and Engineering |
| O/P | |

Your organisation may have a checklist of items that should be included in a design specification. This will ensure that each project produces a complete and comprehensive specification and will help in minimising failures of the safety function due to specification errors.

Lifecycle Phase 4 [Figure 5] may be adequately addressed in a single Functional Design Specification (FDS) or similar document, which sets the scene, defines the process, the environmental and operational considerations and establishes the scope of the following phases.

FIGURE 5. SAFETY LIFECYCLE PHASE 4

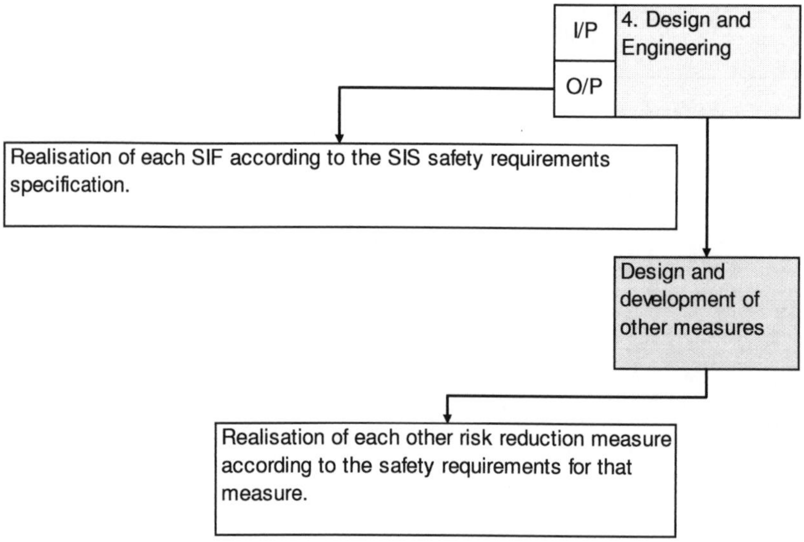

Safety Lifecycle Phases 5 and 6

Lifecycle Phase 5 and Lifecycle Phase 6 identify requirements for SIS installation, commissioning, validation, operation and maintenance, Figure 6.

FIGURE 6. SAFETY LIFECYCLE PHASES 5 AND 6

A plan for the installation and commissioning of the SIS.
Provides planning for the installation and commissioning activities; procedures, techniques and measures to be used; schedule and personnel and departments responsible.

A plan for the overall safety validation of the SIS
Provides planning for the SIS safety validation against the SRS and other reference information i.e. cause and effects charts. Validation will include all relevant modes of operation (start-up, shutdown, maintenance, abnormal conditions etc.), procedures, techniques and measures to be used, schedule, personnel and departments responsible. Will also include validation planning for the safety application software.

Realisation of each SIF according to the SIS safety requirements specification.

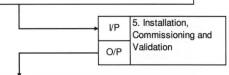

| I/P | 5. Installation, Commissioning and Validation |
| O/P | |

Fully installed and commissioned SIS:
Document installation;
Reference to failure reports;
Resolution of failures.

Confirmation that the SIS meets the specification for the overall safety requirements in terms of the SIF requirements and the safety integrity requirements, taking into account the safety requirements allocation.
Documentation requirements include: chronological validation activities;
version of the safety requirements; safety function being validated;
tools and equipment; results; item under test, procedure applied and test environment; discrepancies; Decisions taken as a result.

A plan for operating and maintaining the SIS
Provides planning for routine and abnormal operation activities; proof testing, maintenance activities, procedures, techniques and measures to be used, schedule, personnel and departments responsible, method of verification against the operation and maintenance procedures.

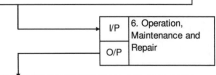

| I/P | 6. Operation, Maintenance and Repair |
| O/P | |

Continuing achievement of the required functional safety for the SIS.
The following shall be implemented:
O&M Plan;
Operation, maintenance and repair procedures.
Implementation of procedures;
Following of maintenance schedules;
Maintain documentation;
Carry out regular FS Audits;
Document modifications.
Chronological documentation of operation and maintenance of the SIS.

Safety Lifecycle Phases 7 and 8

The inputs, outputs and activities associated with Lifecycle Phase 7: Modification, are essentially the same for Lifecycle Phase 8: Decommissioning. In effect, decommissioning is a modification which occurs at the end of the lifecycle and is initiated with the same controls and managed with the same safeguards.

FIGURE 7. SAFETY LIFECYCLE PHASES 7 AND 8

Continuing achievement of the required functional safety for the SIS.
The following shall be implemented:
O&M Plan;
Operation, maintenance and repair procedures.
Implementation of procedures;
Following of maintenance schedules;
Maintain documentation;
Carry out regular FS Audits;
Document modifications.
Chronological documentation of operation and maintenance of the SIS.

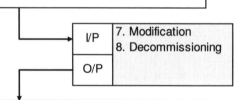

I/P	7. Modification
	8. Decommissioning
O/P	

Achievement of the required functional safety for the SIS, both during and after the modification phase has been maintained. Modification shall only be initiated following authorised request under procedure for FS Management. Request shall include: the hazards that may be affected;
proposed change (hardware and software); reason for change.
Impact analysis shall be carried out.
Chronological documentation of operation and maintenance of the SIS.

3.   HAZARDS AND HAZARD IDENTIFICATION

Lifecycle Phases

Figure 8 shows the phase of the lifecycle that applies.

FIGURE 8. LIFECYCLE PHASE 1

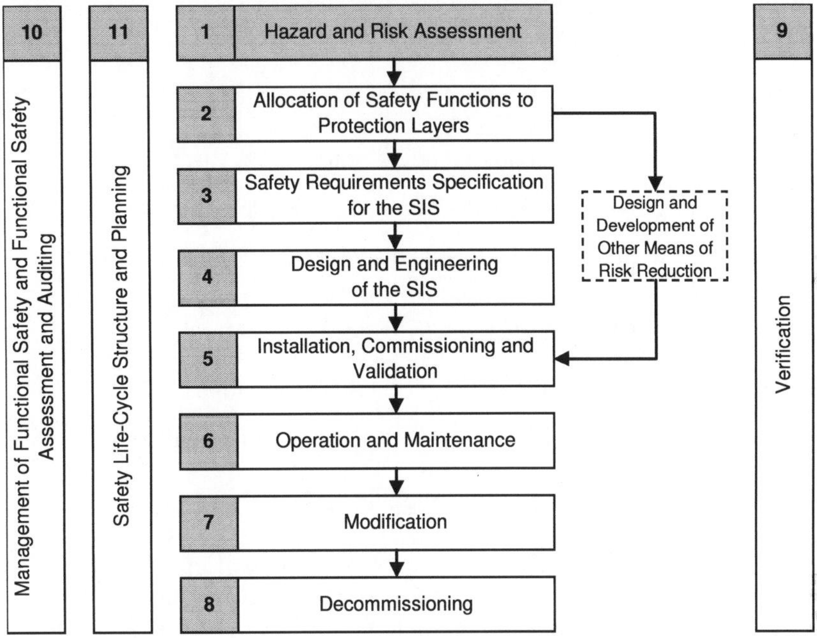

The objective of this Lifecycle Phase 1, as defined in IEC61511-1, 8.1 is to determine:

- Hazards or Hazardous Events of the Process and associated equipment, the sequence of events that lead to the hazard and the process risks involved;

- Requirements for Risk Reduction [5 and 6];

- Safety Functions required to achieve the necessary Risk Reduction [7 and 8].

Hazards

The meaning of the word hazard can be confusing. Often dictionaries do not give specific definitions, or they combine it with the term "risk" for example, as "a danger or risk" which helps explain why many people use the terms interchangeably.

In the context of functional safety, hazards are events which have the potential to cause harm such as personal injury, damage to the environment or the business.

Examples of hazards in the home include:

- Broken glass because it could cause cuts;
- Pools of water because it could cause slips and falls;
- Too many plugs in a socket could overload it and cause a fire.

Examples of hazards at work might include:

- Loud noise because it can cause hearing loss;
- Breathing in asbestos dust because it can cause cancer.

Hazards in the process industry might include:

- The level of liquid in a vessel: a high level may result in an overflow of liquid into gas streams, or an overspill of a dangerous chemical or flammable liquid; a low level may result in dry running of pumps, or gas blowby into downstream vessels.
- The pressure of liquid in a vessel: high pressure may result in loss of containment, leaks or vessel rupture.

The first step in assessing risk is to identify the hazards. There are a number of techniques used for identifying hazards but the technique in most common use is the Hazard and Operability (HAZOP) Study.

Use of HAZOPs in Industry

HAZOPs were originally developed in the UK, by ICI following the Flixborough disaster in 1974, and began to be more widely used within the process industry as a result.

On Saturday 1 June 1974 the Nypro (UK) site at Flixborough was severely damaged by a large explosion that killed 28 workers and injured a further 36. It was recognised that the number of casualties would have been more if the incident had occurred on a weekday, as the main office block was not occupied.

There were 53 reported third party injuries offsite and property in the surrounding area was also damaged. The 18 fatalities in the control room were as a result of the windows shattering and the collapse of the roof. No one escaped. The fires burned for several days and hampered rescue work for the next ten days.

From the chemical industry, through the general exchange of ideas and personnel, HAZOPs were subsequently adopted by the petroleum industry, which has a similar potential for major disasters. They were then taken up by the food and water industries, where the hazard potential is as great, but more concerned with contamination issues rather than explosions or chemical releases.

Reasons to use HAZOPs

Although the design of plant relies upon the application of codes and standards, the HAZOP process allowed the opportunity to supplement these with an imaginative anticipation of the deviations which may occur because of, for example, process conditions or upsets, equipment malfunction or operator error.

In addition, the pressures of project schedules can result in errors or oversights and the HAZOP allows these to be corrected before such changes become too expensive. Because they are easy to understand and can be adapted to any process or business, HAZOPs have become the most widely used hazard identification methodology.

Deviation from Design Intent

All processes, equipment under control or industrial plant have a design intent. This might be to achieve a target production capacity in terms of an annual tonnage of a particular chemical, or a specified number of manufactured items. However, an important secondary design intent may be to operate the process in a safe and efficient manner and to do that, each item of equipment will be required to function effectively. It is this aspect which could be considered the design intent for that particular item of equipment.

For example, as part of our plant production requirement we may need a cooling water facility containing a cooling water circuit with a circulating pump and heat exchanger, as illustrated in Figure 9.

FIGURE 9. DESIGN INTENT

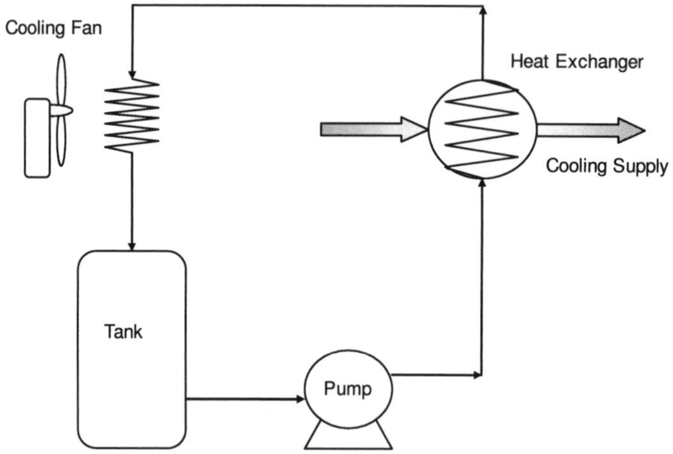

The design intent of this small section of the plant might be to continuously circulate cooling water at a temperature of $x^{\circ}C$ and at a rate of xxx litres per hour. It is usually at this low level of design intent that a HAZOP Study is directed. The use of the word deviation now becomes easier to understand. A deviation or departure from the design intent in the case of our cooling facility would be a reduction of circulating flow, or an increase in water temperature.

Note the difference between a deviation and its cause. In the case above, failure of the pump would be a cause, not a deviation. In this example, an increase in water temperature would be the hazard as it would have the potential to cause harm such as personal injury, damage to the environment or the business.

HAZOP Technique

HAZOPs are used to identify potential hazards and operability problems caused by deviations from the design intent of both new and existing process plant and are generally carried out periodically throughout the plant's life. Certainly an initial or preliminary HAZOP should be carried out early in the design phase. The process should be reviewed as the development progresses and whenever major modifications are proposed, and finally at the end of the development to ensure there are no residual risks prior to the build stage.

A HAZOP is conducted in a meeting forum between interested parties with sufficient knowledge and experience of the operation and maintenance of the plant. The meeting is a structured brainstorming session whereby guidewords are employed to stimulate ideas about what the hazards could be. The minutes of the meeting record the discussion and capture information about potential hazards, their causes and consequences.

HAZOP Study Team

It is important that a HAZOP team is made up of personnel who will bring the best balance of knowledge and experience, of the type of plant being considered, to the study. A typical HAZOP team is made up as follows:

Name	Role
Chairman	To explain the HAZOP process, to direct discussions and facilitate the HAZOP. Someone experienced in HAZOP but not directly involved in the design, to ensure that the method is followed carefully.
Secretary	To capture the discussion of the HAZOP Meeting and provide a visible record of the discussions. To log recommendations or actions.
Process Engineer	Usually the chemical engineer responsible for the process flow diagram and development of the Piping and Instrumentation Diagrams (P&IDs).
User / Operator	To advise on the use and operability of the process, and the effect of deviations.
C&I Specialist	Someone with relevant technical knowledge of Control and Instrumentation.
Maintainer	Someone concerned with maintenance of the process.
A design team representative	To advise on any design details or provide further information.

Information Used in the HAZOP

The following items may be available to view by the HAZOP team:

- Piping and Instrumentation Diagrams (P&IDs) for the facility;
- Process Description or Philosophy Documents;
- Existing Operating and Maintenance Procedures;
- Cause and Effects (C&E) charts;
- Plant layout drawings.

The HAZOP Procedure

The HAZOP procedure involves taking a full description of the process and systematically questioning every part of it to establish how deviations from the design intent can have a negative effect upon the safe and efficient operation of the plant.

The procedure is applied in a structured way by the HAZOP team, and it relies upon them releasing their imagination in an effort to identify credible hazards. In practice, many of the hazards will be obvious, such as an increase in temperature, but the strength of the technique lies in its ability to discover less obvious hazards, however unlikely they may seem at first consideration.

Guidewords

The HAZOP process uses guidewords to focus the attention of the team upon deviations of the design intent, their possible causes and consequences. These Guidewords are divided into two sub-sets:

Primary Guidewords which focus attention upon a particular aspect of the design intent or an associated process condition or parameter i.e. flow, temperature, pressure, level etc.;

Secondary Guidewords which, when combined with a primary guideword, suggest possible deviations i.e. more temperature, less level, no pressure, reverse flow etc.

The entire technique depends upon the effective use of these guidewords, so their meaning and use must be clearly understood by the team.

It should be noted that the use of guidewords is simply to stimulate the imagination into what could happen. Not all guidewords will be meaningful, not all hazards will be credible. In these cases, it is recommended that where the team identify meaningless or incredible events, then these are recorded as such and the team waste no time in moving on.

Modes of Operation

As a HAZOP is a hazard and operability study, it is important to consider not only the normal operation of the process but also other abnormal modes, such as start-up, shutdown, filling, emptying, by-pass, proof test.

This may be accomplished by considering each operational mode specified in the scope, as a separate exercise and producing separate HAZOP analyses for each. Alternatively, for relatively simple systems, an additional column can be included in the worksheets to identify the mode. A single HAZOP analysis can thus consider all operational modes.

Recording the HAZOP

There are software tools available to guide you through the HAZOP process. Alternatively, a simple spreadsheet can be constructed to record the discussions and findings. Spreadsheets allow for easy sorting and categorisation, and they also provide visibility and traceability between entries so that cross-referencing with other analyses can be maintained.

It is recommended to record every event and guideword combination considered. Where applicable, it can be noted: No Credible Cause, No Consequence, or No Hazard. This is classified as Full Recording, and it results in a HAZOP Report which demonstrates that a comprehensive and rigorous study has been undertaken. This will be invaluable in the assessment of safety and operability of later plant modifications.

In addition to the above, the Secondary words 'All' and 'Remainder' are often used. For example, some combinations of Primary Guideword may be identified as having credible causes e.g. Flow/No, Flow/Reverse. For other combinations (Flow/Less, Flow/More, Flow/Other), where no credible causes can be identified, the combination 'Flow/Remainder' can be used.

Identifying Hazards – HAZOP Worksheet Headings

The following table presents an example HAZOP Worksheet for a Decompression Chamber. Note that this is purely representational, and not intended to illustrate an actual system.

Reference

It is always worth including a reference column so that each entry can be referred to from other analyses and also allows traceability to subsequent analysis, e.g. LOPA [8].

Guidewords

Primary and secondary guidewords should be used. The internet can provide various lists of guidewords that apply to different businesses and industries.

Deviation

The deviation is the departure from the design intent prompted by the primary and secondary guidewords and represents the identified hazard.

Cause

Potential causes which would result in the deviation occurring. It is important to include specific information about the cause for example, entry 1.06 in the HAZOP example, is concerned with an increase in oxygen concentration caused by an O_2 sensor failure. The sensor could fail in a number of ways but only reading a false low O_2 concentration would result in the hazardous condition.

Consequence

The consequences that would arise from the effect of the deviation and, if appropriate, from the cause itself. Always be explicit in recording the consequences. Do not assume that the reader at some later date will understand what the hazard is or how the consequences will develop.

When assessing the consequences, it is important not to take any credit for protective systems or instruments which are already included in the design.

Ref	Primary Keyword	Secondary Keyword	Deviation	Hazard	Cause	Consequence	Safeguards
01.01	Flow	More	Increase in airflow into the chamber	Rate of increase in chamber pressure excessive.	PCV02 fails open	Increase in chamber pressure. Squeeze effects. May affect eyes, face, teeth and sinuses.	Dive Control Manager can isolate the chamber.
01.02					ADCS failure demands PCV02 open	Risk of punctured ear drum.	Chamber occupants can isolate the chamber.
01.03					PT fails and reads low pressure	Marginal (single severe or number of minor injuries).	
01.04			Increase in oxygen flow into the chamber	Elevated ppO2 within chamber.	PCV01 fails open	Pulmonary oxygen poisoning or possibly Hyperoxia leading to seizure or convulsion due to Central Nervous System (CNS) oxygen toxicity.	Dive Control Manager can isolate the chamber.
01.05					ADCS failure demands PCV01 open	Catastrophic (more than 1 death).	Chamber occupants can isolate the chamber.
01.06					O2 sensor fails and reads low O2 concentration		
01.07		Less	Reduction of airflow into the chamber	Unintended significant chamber pressure loss.	PCV02 fails closed	Rapid loss of pressure. Inadvertent decompression of divers. Potential for decompression sickness or illness (the bends), embolism and other pressure effects of ascent.	Dive Control Manager can isolate the chamber.
01.08					ADCS failure demands PCV02 closed		Chamber occupants can isolate the chamber.
01.09					PT fails and reads high pressure	Catastrophic (more than 1 death).	
01.10			Reduction of oxygen flow into the chamber	Decreased ppO2 within Chamber.	PCV01 fails closed	Hypoxia leading to unconsciousness and death.	Dive Control Manager can isolate the chamber.
01.11					ADCS failure demands PCV01 closed	Catastrophic (more than 1 death).	Chamber occupants can isolate the chamber.
01.12					O2 sensor fails and reads high O2 concentration		

When documenting consequences, it is important to remember that the HAZOP may be used to determine risk and therefore a full and complete description of how the hazard may develop and result in consequences is essential. For example, consequences may be described as:

"Potential overpressure leading to rupture of gas discharge pipework and loss of containment. Large volume gas release ignites on hot machine exhaust resulting in explosion or flash fire with potential fatalities of up to two maintenance personnel. Compressor damage of up to £2 million and loss of production for up to 1 year"

Safeguards

Any existing protective devices which either prevent the cause or safeguard against the consequences would be recorded in this column. Safeguards need not be restricted to hardware, where appropriate, credit can be taken for procedural aspects such as regular plant inspections (if you are sure that they will actually be carried out AND that they can either prevent or safeguard).

Action

The Action column provides an opportunity to make recommendations either to initiate some corrective action, such as investigating what additional safeguards can be implemented. Possible actions fall into two groups:

- Actions that eliminate the cause;
- Actions that mitigate the consequences.

Eliminating the cause of the hazard is always the preferred solution. Only when this is not feasible, should consideration be given to mitigating the consequences.

Example HAZOP: Separator Vessel

The following example shows a simplified schematic of a process separator vessel. The vessel takes in the process liquid which is heated by a gas burner. Vapour is separated from the process liquid and released for export. The remaining concentrated liquid is drawn off from the bottom of the vessel when the reaction is complete, Figure 10.

The vessel has a Distributed Control System (DCS) which controls liquid level within the vessel, gas pressure and temperature.

FIGURE 10. SEPARATOR VESSEL

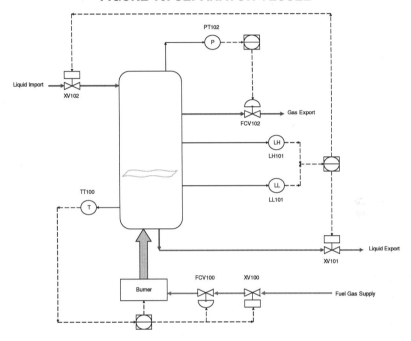

An example HAZOP for this separator vessel, is shown in the following pages.

Each entry in the HAZOP is uniquely referenced so that it can be referred to from other analyses and allowing traceability to subsequent analysis, e.g. LOPA [8]. The primary and secondary guidewords provide assistance in developing deviations from the design intent in order to identify potential hazards. Possible causes which would result in the deviation occurring have been identified and the consequences that would arise from the effect of the deviation have been recorded.

Any recommendations resulting from the HAZOP, either to initiate some corrective action, such as investigating what additional safeguards can be implemented have been documented in the form of an action list.

The following worksheet is to be read across as a double page.

Separator Vessel HAZOP

Ref	Primary Keyword	Secondary Keyword	Deviation	Hazard
01.01	Flow	More	High flow of process liquid into vessel.	High flow into vessel could result in high level, liquid carry over into gas export.
01.02			High flow of process liquid out from vessel liquid export.	High flow from vessel could result in low level, gas blow-by into liquid export.
01.03			High flow of gas out from vessel gas export.	No credible hazard
01.04		Less	Low flow of process liquid into vessel.	Low flow into vessel could result in low level, gas blow-by into liquid export.
01.05			Low flow of process liquid out from vessel liquid export.	Low flow from vessel could result in high level, liquid carry over into gas export.
01.06			Low flow of gas out from vessel gas export.	No credible hazard
01.07		Reverse	Not credible.	No credible hazard
01.08		Also	Not credible.	No credible hazard
01.09		Other	Not credible.	No credible hazard
01.10	Pressure	More	High pressure in vessel.	Vessel rupture and gas release.
01.11		Less	Low pressure in vessel.	Vessel rupture and gas release.
01.12		Reverse	Not credible.	No credible hazard
01.13		Also	Not credible.	No credible hazard
01.14		Other	Not credible.	No credible hazard
01.15	Temperature	More	High temperature in vessel.	High temperature leads to high pressure, vessel rupture and gas release.
01.16		Less	Low temperature in vessel.	Potential liquid freezing (solifiying), vessel rupture and loss of containment.
01.17		Reverse	Not credible.	No credible hazard
01.18		Also	Not credible.	No credible hazard
01.19		Other	Not credible.	No credible hazard
01.20	Level	More	High level in vessel.	High level in vessel could result in liquid carry over into gas export.
01.21		Less	Low level in vessel.	Low level in vessel could result in gas blow-by into liquid export.
01.22		Reverse	Not credible.	No credible hazard
01.23		Also	Not credible.	No credible hazard
01.24		Other	Not credible.	No credible hazard

Separator Vessel HAZOP (continued)

Consequence	Safeguards	Action
Equipment damage downstream requiring vessel replacement estimated at £10M and process shutdown for 6 months.	Level control.	Consider installation of high level alarm.
Equipment damage downstream requiring vessel cleaning estimated at £2M and process shutdown for 6 weeks.	Level control.	Consider installation of low level alarm.
None.	None.	
Equipment damage downstream requiring vessel cleaning estimated at £2M and process shutdown for 6 weeks.	Level control.	Consider installation of low level alarm.
Equipment damage downstream requiring vessel replacement estimated at £10M and process shutdown for 6 months.	Level control.	Consider installation of high level alarm.
None.	None.	
None.	None.	
None.	None.	
None.	None.	
Gas release ignites on burner and hot surfaces. Possibly two maintainer fatalities. Equipment damage requiring vessel replacement estimated at £10M and processs shutdown for 1 year. Minor environmental release.	Pressure control.	Consider installation of high level alarm.
Gas release ignites on burner and hot surfaces. Possibly two maintainer fatalities. Equipment damage requiring vessel replacement estimated at £10M and processs shutdown for 1 year. Minor environmental release.	Pressure control.	Consider installation of low level alarm.
None.	None.	
None.	None.	
None.	None.	
Gas release ignites on burner and hot surfaces. Possibly two maintainer fatalities. Equipment damage requiring vessel replacement estimated at £10M and processs shutdown for 1 year. Minor environmental release.	Temperature control.	Consider installation of high temperature alarm.
Equipment damage requiring vessel replacement estimated at £10M and process shutdown for 6 months. Environmental release requiring notification order.	Temperature control.	Consider installation of low temperature alarm.
None.	None.	
None.	None.	
None.	None.	
Equipment damage downstream requiring vessel replacement estimated at £10M and process shutdown for 6 months.	Level control.	Consider installation of high level alarm.
Equipment damage downstream requiring vessel cleaning estimated at £2M and process shutdown for 6 weeks.	Level control.	Consider installation of low level alarm.
None.	None.	
None.	None.	
None.	None.	

HAZOP Results

Summarising, the hazards identified are:

Hazard	Consequence
High level in vessel could result in liquid carry over into gas export.	Equipment damage downstream requiring vessel replacement estimated at £10M and process shutdown for 6 months.
High pressure causes vessel rupture and gas release.	Gas release ignites on burner and hot surfaces. Possibly two maintainer fatalities. Equipment damage requiring vessel replacement estimated at £10M and process shutdown for 1 year. Minor environmental release.
High temperature leads to high pressure, vessel rupture and gas release.	Gas release ignites on burner and hot surfaces. Possibly two maintainer fatalities. Equipment damage requiring vessel replacement estimated at £10M and process shutdown for 1 year. Minor environmental release.
Low level in vessel could result in gas blow-by into liquid export.	Equipment damage downstream requiring vessel cleaning estimated at £2M and process shutdown for 6 weeks.
Low pressure causes vessel rupture and gas release.	Gas release ignites on burner and hot surfaces. Possibly two maintainer fatalities. Equipment damage requiring vessel replacement estimated at £10M and process shutdown for 1 year. Minor environmental release.
Low temperature, potential liquid freezing (solidifying), vessel rupture and loss of containment.	Equipment damage requiring vessel replacement estimated at £10M and process shutdown for 6 months. Environmental release requiring notification order.

The list of identified hazards forms a Hazard Log for the system. The hazard log should remain a live document throughout the system lifecycle and can be added to, or revised as other studies are completed.

Each of the identified hazards could have possible safety, environmental or commercial consequences but in order to address our obligations under the Health and Safety at Work Act [20.3], we must determine the level of risk associated with each of the hazards, [4].

4. RISK AND RISK REDUCTION

Concept of Risk

A risk is the likelihood that a hazard will cause a measureable adverse effect. It is therefore a two-part concept and you have to have both parts to make sense of it. Likelihoods can be expressed in different ways, for example as a probability: one in a thousand; as a frequency or rate: 1000 cases per year, or in a qualitative way: negligible or significant.

The effect can be described in many different ways. For example:

- A single employee serious injury or fatality;
- Multiple third-party injuries;
- Members of the public exposed to toxic gas;

The annual risk of an employee experiencing a fatal accident [effect] at work from contact with moving machinery [hazard] is less than one in 100,000 [likelihood].

Risk therefore needs to be quantified in two dimensions. The impact, or the consequences of the hazard needs to be assessed, and the probability of occurrence needs to be evaluated. For simplicity, you could rate each on a 1 to 4 scale, as shown in Figure 11 where the larger the number, the greater the impact or probability of occurrence. As a general principle, by using a risk matrix such as this, a priority can be established and the risk evaluated.

FIGURE 11. RISK MATRIX

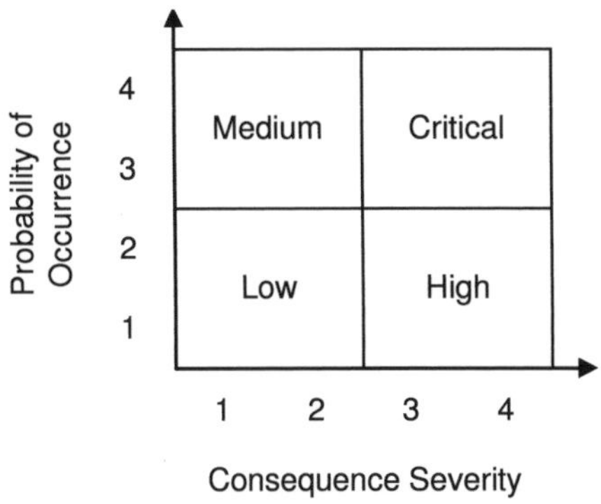

Probability of Occurrence

Consequence Severity

If the probability of occurrence is high, and the consequence severity is low, this might represent a Medium risk. On the other hand if the consequence severity is high, and probability of occurrence is low, the risk might be considered High. Typically, a remote chance of a catastrophic event should warrant more attention than a minor nuisance which occurs frequently.

So far, risks to personal safety, to the environment, or to the business in terms of asset damage or revenue earing capacity have been discussed but there is no reason why the same approach cannot be adopted for other risks. For example, utility companies may wish to address security of supply risks. Similarly, where a business has experienced a major accident in the past, then the consequences of another similar event in the future, and the potential for national or international press coverage may be so severe that risks to the company's reputation may need to be considered.

Separator Vessel HAZAN

A first pass assessment of risk can usually be performed as part of the HAZOP and this is known as a Hazard Analysis (HAZAN). As in Figure 11, each hazard can be categorised in terms of its severity (in this case 1 to 4, where 4 is the most severe) and probability of occurrence, or frequency (again 1 to 4, where 4 is the most likely).

The Separator Vessel HAZOP example can be developed and multiplying the severity and frequency categories together gives a preliminary measure of risk in the form of a Risk Priority Number (RPN) which can be used to prioritise risk reduction actions, Table 1.

TABLE 1. SEPARATOR VESSEL HAZAN

Ref	Consequence	Sev Cat	Freq Cat	RPN
01.01	Equipment damage downstream requiring vessel replacement estimated at £10M and process shutdown for 6 months.	3	2	6
01.02	Equipment damage downstream requiring vessel cleaning estimated at £2M and process shutdown for 6 weeks.	2	1	2
01.03	None.			

Ref	Consequence	Sev Cat	Freq Cat	RPN
01.04	Equipment damage downstream requiring vessel cleaning estimated at £2M and process shutdown for 6 weeks.	2	2	4
01.05	Equipment damage downstream requiring vessel replacement estimated at £10M and process shutdown for 6 months.	3	1	3
01.06	None.			
01.07	None.			
01.08	None.			
01.09	None.			
01.10	Gas release ignites on burner and hot surfaces. Possibly two maintainer fatalities. Equipment damage requiring vessel replacement estimated at £10M and process shutdown for 1 year. Minor environmental release.	4	2	8
01.11	Gas release ignites on burner and hot surfaces. Possibly two maintainer fatalities. Equipment damage requiring vessel replacement estimated at £10M and process shutdown for 1 year. Minor environmental release.	4	1	4
01.12	None.			
01.13	None.			
01.14	None.			
01.15	Gas release ignites on burner and hot surfaces. Possibly two maintainer fatalities. Equipment damage requiring vessel replacement estimated at £10M and process shutdown for 1 year. Minor environmental release.	4	1	4
01.16	Equipment damage requiring vessel replacement estimated at £10M and process shutdown for 6 months. Environmental release requiring notification order.	3	1	3
01.17	None.			
01.18	None.			
01.19	None.			

Ref	Consequence	Sev Cat	Freq Cat	RPN
01.20	Equipment damage downstream requiring vessel replacement estimated at £10M and process shutdown for 6 months.	3	2	6
01.21	Equipment damage downstream requiring vessel cleaning estimated at £2M and process shutdown for 6 weeks.	2	1	2
01.22	None.			
01.23	None.			
01.24	None.			

HAZOP Actions

The HAZOP Worksheets also identify actions for further investigation. The Action column provides an opportunity to make recommendations either to initiate some corrective action, such as investigating what additional safeguards can be implemented.

Possible actions fall into two groups:

- Actions that eliminate the cause;

- Actions that mitigate the consequences.

Eliminating the cause of the hazard is always the preferred solution. Only when this is not feasible, should consideration be given to mitigating the consequences. In this example, the following actions were identified:

Ref	Hazard	Consequence	Action	Action allocated	Completion date
01.01	High flow into vessel could result in high level, liquid carry over into gas export.	Equipment damage downstream.	Consider installation of high level alarm.	S Smith C&I Dept	14 Apr 12

Ref	Hazard	Consequence	Action	Action allocated	Completion date
01.02	High flow from vessel could result in low level, gas blow-by into liquid export.	Equipment damage downstream.	Consider installation of low level alarm.	S Smith C&I Dept	14 Apr 12
01.04	Low flow into vessel could result in low level, gas blow-by into liquid export.	Equipment damage downstream.	Consider installation of low level alarm.	S Smith C&I Dept	14 Apr 12
01.05	Low flow from vessel could result in high level, liquid carry over into gas export.	Equipment damage downstream.	Consider installation of high level alarm.	S Smith C&I Dept	14 Apr 12
01.10	Vessel rupture and gas release.	Possible maintainer fatalities. Equipment damage. Environmental release.	Consider installation of high pressure alarm.	J Jones Process Dept	21 Apr 12
01.11	Vessel rupture and gas release.	Possible maintainer fatalities. Equipment damage. Environmental release.	Consider installation of low pressure alarm.	J Jones Process Dept	21 Apr 12
01.15	High temperature leads to high pressure, vessel rupture and gas release.	Possible maintainer fatalities. Equipment damage. Environmental release.	Consider installation of high temperature alarm.	V White C&I Dept	21 Apr 12
01.16	Potential liquid freezing, vessel rupture and loss of containment.	Equipment damage. Environmental release.	Consider installation of low temperature alarm.	V White C&I Dept	21 Apr 12

Ref	Hazard	Consequence	Action	Action allocated	Completion date
01.20	High level in vessel could result in liquid carry over into gas export.	Equipment damage downstream.	Consider installation of high level alarm.	S Smith C&I Dept	14 Apr 12
01.21	Low level in vessel could result in gas blow-by into liquid export.	Equipment damage downstream.	Consider installation of low level alarm.	S Smith C&I Dept	14 Apr 12

Examples of Risk Matrix Categorisations

Figure 12 presents similar information to the simple risk matrix above. The severity of the consequences has been classified by simple generic descriptions, e.g. incidental, minor, severe, catastrophic etc. If a HAZOP has been conducted, then the likely consequences of the hazards identified will probably be known and these can be grouped and categorised.

Quantification of the likelihood of occurrence is more difficult.

Figure 12 shows one approach whereby the likelihood is categorised descriptively from very frequent, e.g. the hazard occurs several times a year on site, to very rare, e.g. never heard of in the industry, or never heard of in any industry. With such a qualitative description of the likelihood of occurrence it is possible to assign ranges of frequencies to each category.

The resulting table thus allows a categorisation of risks ranging from very low (VL), to low (L), medium (M), high (H) and very high (VH) according to the severity category and the frequency.

FIGURE 12. RISK MATRIX

Severity		A — Never heard of in any industry / work type (<10^{-6}/yr)	B — Never heard of in industry / work type (10^{-6}-10^{-5}/yr)	C — Heard of in industry / work type (10^{-5}-10^{-4}/yr)	D — Occurred within business (10^{-4}-10^{-3}/yr)	E — Occurs several times within business (10^{-3}-10^{-2}/yr)	F — Occurs on site (10^{-2}-10^{-1}/yr)	G — Occurs several times on site (10^{-1}-1/yr)	H — Occurs several times a year on site (>1/yr)
Catastrophic 10^{-6}/yr	6	VL	L	M	H	VH	VH	VH	VH
Severe 10^{-5}/yr	5		VL	L	M	H	VH	VH	VH
Major 10^{-4}/yr	4			VL	L	M	H	VH	VH
Moderate 10^{-3}/yr	3				VL	L	M	H	VH
Minor 10^{-2}/yr	2					VL	L	M	H
Incidental 10^{-1}/yr	1						VL	L	M

Likelihood

Risk Quantification

The tolerability of risk has so far been qualitative. The quantification of the tolerability of risks to personal safety, depends on how risks are perceived and several factors can influence this including:

- personal experience of adverse effects;
- social or cultural background and beliefs;
- the degree of control one has over a particular risk;
- the extent to which information is gained from different sources e.g. the media.

Clearly there risks that are so high, they are obviously unacceptable, for example smoking while pregnant, and others where the risk is so low as to be negligible, such as boiling a saucepan of milk. Of course, the most interesting area for discussion is in the grey tolerable risk area in between. The task is therefore to define the two boundary conditions:

- between unacceptable risk and tolerable risk, and;
- between tolerable and acceptable risk.

The HSE guidance document Reducing Risks, Protecting People (R2P2) [20.5] proposes that an individual **risk of death of one in a million, per year, for both employees and members of the public** corresponds to a very low level of risk and should be used as the broadly acceptable (negligible) risk boundary.

R2P2 goes on to suggest that an **individual risk of death of 1 in 1,000 per annum** should represent the boundary condition between what is just tolerable for a substantial category of workers for a large part of their working lives, and what is unacceptable for any but fairly exceptional groups. In the UK, the occupational health and safety target is to achieve a level to which nearly all of the population could be exposed day after day, without adverse effects.

For members of the public who have risk imposed upon them, this limit is judged to be an order of magnitude lower at **1 in 10,000 per annum**.

The criteria adopted by the HSE can be demonstrated in a framework known as the Tolerability of Risk (TOR), Figure 13. The maximum tolerable individual risk and broadly acceptable risk criteria have been marked.

FIGURE 13. TOLERABILITY OF RISK

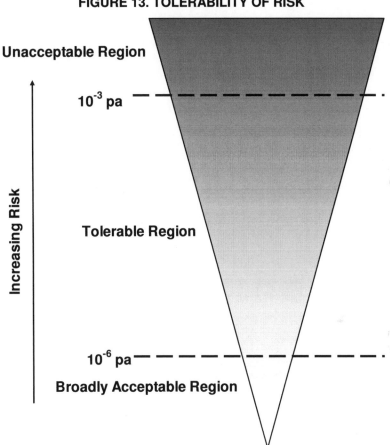

Tolerability and Acceptability of Risk

In determining quantitative risk posed by hazards identified in say, a HAZOP, it is necessary to set quantitative risk criteria and to take account of other occupational hazards that an individual will be exposed to during the working day. It is not unreasonable to make the assumption that an individual will be exposed to an estimated 10 such hazards. The tolerability of risk criteria, can then be apportioned between these 10 hazards giving a maximum tolerable individual risk of death of 1 in 10,000 per annum, Figure 14.

FIGURE 14. INDIVIDUAL RISK CRITERIA

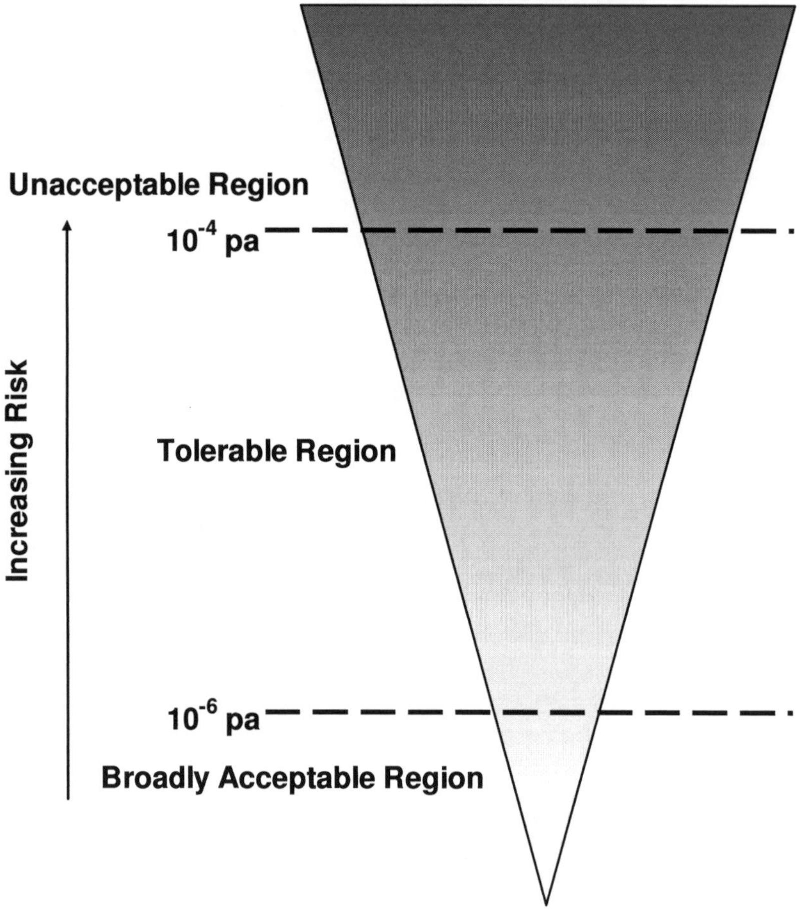

Unacceptable Region

10^{-4} pa

Increasing Risk

Tolerable Region

10^{-6} pa

Broadly Acceptable Region

The broadly acceptable risk boundary for individual risk of death for both employees and members of the public remains at one in a million per year since this is already considered negligible.

From the maximum tolerable individual risk of death of 1 in 10,000 per annum, other maximum tolerable risk values can be determined by adjusting the maximum tolerable frequency for each consequence depending on severity and whether or not third parties are involved, Figure 15.

FIGURE 15. TOLERABILITY OF RISK SUMMARY

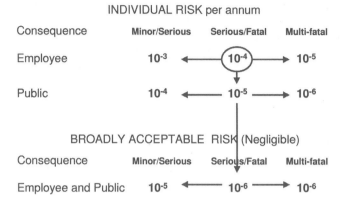

INDIVIDUAL RISK per annum

Consequence	Minor/Serious	Serious/Fatal	Multi-fatal
Employee	10^{-3} ⟵	(10^{-4}) ⟶	10^{-5}
Public	10^{-4} ⟵	10^{-5} ⟶	10^{-6}

BROADLY ACCEPTABLE RISK (Negligible)

Consequence	Minor/Serious	Serious/Fatal	Multi-fatal
Employee and Public	10^{-5} ⟵	10^{-6} ⟶	10^{-6}

Tolerability of Risk

The tolerability of risk summary [Figure 15] can be represented graphically as shown in [Figure 16]. The frequency axis does not show values lower than 10^{-6} /yr as this is considered negligible risk whatever the consequences.

FIGURE 16. TOLERABILITY OF RISK SUMMARY

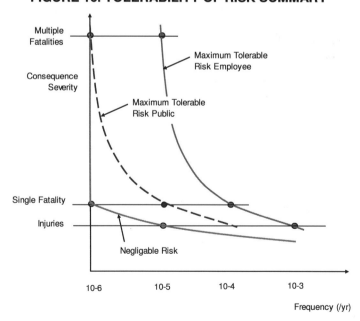

Requirements for Compliance

The requirements for safety integrity levels are derived from the likely frequencies of hazardous events. Depending upon the consequences of a hazard, a maximum tolerable frequency will be determined and a safety function engineered to bring the frequency down to a tolerable level.

The risk reduction required of the safety function provides the first requirement for compliance with the standard: this is the numerical reliability measure.

The numerical reliability measure is categorised by value, into bands or safety integrity levels (SILs). There are four SILs based on the target reliability measure required. SIL4 provides the highest level of integrity, the greatest amount of risk reduction and most onerous reliability target. SIL1 provides the lowest level of integrity and least onerous reliability target.

The ALARP Principle

The above overview of hazard and risk analysis illustrates how the process risks may be determined and the maximum tolerable risk achieved. However, under the HSAWA, additional efforts must continue to be made to reduce the risk further, by other means, until it can be shown that the risk is 'As Low As Reasonably Practicable' (ALARP), i.e. further risk reduction is not cost-effective, [1].

5. THE ALARP PRINCIPLE

Benefits and Sacrifices

Using "reasonably practicable" sets goals for duty-holders, rather than being prescriptive. This flexibility is a great advantage in that it allows duty-holders to choose the method that is best for them and so it supports innovation, but it has its drawbacks, too. Deciding whether a risk is ALARP can be challenging because it requires duty-holders, and assessors, to exercise judgment.

The main tests that are applied in regulating industrial risks involve determining whether:

a) the risk is so great that it must be refused altogether;

b) the risk is, or has been made, so small as to be negligible;

c) the risk falls between the two states specified in a) and b) above and has been reduced to a level which is 'As Low As Reasonably Practicable'.

'Reasonably practicable' is difficult to quantify. It implies that a computation must be made in which the additional risk reduction that may be achieved, is balanced against the sacrifice involved (in money, time or trouble) in achieving it. If there is gross disproportion between them: the benefit being insignificant in relation to the cost; then the risk is considered ALARP. Thus, demonstration that risks have been reduced ALARP involves an assessment of:

*   the risk to be avoided;

*   the sacrifice (in money, time and trouble) involved in taking measures to avoid that risk;

*   a comparison of the two.

This process can involve varying degrees of rigour which will depend upon:

*   the nature of the hazard;

*   the extent of the risk;

*   the control measures to be adopted.

However, duty-holders (and regulators) should not be overburdened if such rigour is not warranted. The greater the initial level of risk under consideration, the greater the degree of rigour is required.

Disproportionality

A Cost Benefit Analysis (CBA) can help a duty holder make judgments on whether further risk reduction measures are justified. Additional risk reduction measures can be considered reasonably practicable unless the costs of implementing the measures are grossly disproportionate to the benefits. Put simply if cost / benefit > DF, where DF is the 'disproportion factor', then the measure can be considered not worth doing for the risk reduction achieved.

DFs that may be considered gross vary from upwards of 1 depending on a number of factors including the severity of the consequences and the frequency of realising those consequences, i.e. the greater the risk, the greater the DF.

What is Gross Disproportion?

HSE has not formulated an algorithm which can be used to determine when the degree of disproportion can be judged as 'gross'. There is no authoritative guidance from the Courts as to what factors should be taken into account in determining whether cost is grossly disproportionate. Therefore a judgment must be made on a case by case basis and some clues or guidance can be obtained from inquiries into major accidents.

From the 1987 Sizewell B Inquiry, the following DFs were used:

- for low risks to members of the public a factor of 2;
- a factor of up to 3 (i.e. costs three times larger than benefits) applied for risks to workers;
- for high risks a factor of 10.

Cost Benefit Analysis (CBA)

For many ALARP decisions, the HSE does not expect duty holders to undertake a detailed Cost Benefit Analysis (CBA), a simple comparison of costs and benefits may suffice.

A CBA should only be used to support ALARP decisions. It should not form the sole argument of an ALARP decision nor be used to undermine existing standards and good practice. A CBA on its own does not constitute an ALARP case and cannot be used to argue against statutory duties, cannot justify risks that are intolerable, nor justify what is evidently poor engineering.

Justifiable costs that may be taken into account in a CBA include:

- Installation;
- Operation;
- Training;
- Any additional maintenance;
- Business losses that would follow from any shutdown undertaken solely for the purpose of putting the measure into place;
- Interest on deferred production, e.g. oil or gas remaining in an oil/gas field while work is carried out on a platform;
- All claimed costs must be those incurred by the duty holder (costs incurred by other parties, e.g. members of the public should not be counted);
- The costs considered should only be those necessary for the purpose of implementing the risk reduction measure (no gold plating or deluxe measures).

Justifiable benefits that may be claimed in a CBA can include all the benefits of implementing a safety improvement measure in full, in that they are not underestimated in any way. The benefits should include all reduction in risk to members of the public, to workers and to the wider community and can include:

- Prevented fatalities;
- Prevented injuries (major to minor);
- Prevented ill health;
- Prevented environmental damage if relevant (e.g. COMAH).

Benefits claimed can also include the avoidance of the deployment of emergency services and avoidance of countermeasures such as evacuation and post-accident decontamination if appropriate. However, in order to compare the benefits of implementing a safety improvement with the associated costs, the comparison must be conducted on a common basis. A simple method for coarse screening of measures puts the costs and benefits into a common format of '£s per year' for the lifetime of a plant.

Table 2 shows some typical monetary values that could be used.

TABLE 2. TYPICAL COURT AWARDS (2003)

Fatality		£1,336,800 (x2 for cancer)
Injury	Permanently incapacitating injury. Some permanent restrictions to leisure and possibly some work activities.	£207,200
	Serious. Some restrictions to work and leisure activities for several weeks or months.	£20,500
	Slight injury involving minor cuts and bruises with a quick and complete recovery.	£300
Illness	Permanently incapacitating illness. Some permanent restrictions to leisure and possibly some work activities.	£193,100
	Other cases of ill health. Over one week absence. No permanent health consequences.	£2,300 + £180 per day absence.
Minor	Up to one week absence. No permanent health consequences.	£50

Example

Question: Consider a chemical plant with a process that if it were to explode could lead to:

- 20 fatalities;
- 40 permanently injured;
- 100 seriously injured;
- 200 slightly injured.

The rate of this explosion happening has been analysed to be about 10^{-5} per year, which is equivalent to 1 in 100,000 per year. The plant has an estimated lifetime of 25 years. How much could the organisation reasonably spend to eliminate the risk from the explosion?

Answer: If the risk of explosion were to be eliminated the benefits can be assessed to be:

Fatalities:	20 x £1,336,800 x 10^{-5} x 25 yrs	=	£6684
Permanent injuries:	40 x £207,200 x 10^{-5} x 25 yrs	=	£2072
Serious injuries:	100 x £20,500 x 10^{-5} x 25 yrs	=	£512
Slight Injuries:	200 x £300 x 10^{-5} x 25 yrs	=	£15
Total benefits		=	**£9,283**

The sum of £9,283 is the estimated benefit of eliminating the major accident explosion at the plant on the basis of avoidance of casualties. This method does not include discounting or take account of inflation.

For a measure to be deemed not reasonably practicable, the cost has to be grossly disproportionate to the benefits. In this case, the DF will reflect that the consequences of such explosions are high. A DF of more than 10 is unlikely and therefore it might be reasonably practicable to spend up to somewhere in the region of £93,000 (£9300 x 10) to eliminate the risk of an explosion. The duty holder would have to justify the use of a smaller DF.

This type of simple analysis can be used to eliminate or include some measures by costing various alternative methods of eliminating or reducing risks.

Alternative approach

It is more likely that a safety improvement measure will not eliminate a risk, just reduce the risk by an amount and we will need to evaluate the risk reduction provided as the benefit against the cost of implementation.

Some organisations operate a cost per life saved target (or Value of Preventing a statistical Fatality: VPF).

The cost of preventing fatalities over the life of the plant is compared with the target VPF. The improvements will be implemented unless costs are grossly disproportionate.

Example

Question: Application of ALARP

A £2M cost per life saved target is used in a particular industry. A maximum tolerable risk target of 10^{-5} pa has been established for a particular hazard, which is likely to cause 2 fatalities.

The proposed safety system has been assessed and a risk of 8.0×10^{-6} pa predicted. Given that the Broadly Acceptable (Negligible) Risk is 10^{-6} pa then the application of ALARP is required.

In this example, for a cost of £10,000, additional instrumentation and redundancy will reduce the risk to 2.0×10^{-6} pa (just above the negligible region) over the life of the plant which is 30 yrs.

Should the proposal be adopted?

Answer: Number of lives saved over the life of the plant is given by:

$$N \quad = \quad \text{(reduction in the fatality frequency) x number of fatalities per incident x life of the plant}$$

$$= \quad (8.0 \times 10^{-6} - 2.0 \times 10^{-6}) \times 2 \times 30$$

$$= \quad 3.6 \times 10^{-4}$$

Hence the cost per life saved is:

$$\text{VPF} \quad = \quad £10000 / 3.6 \times 10^{-4}$$

$$= \quad £27.8M$$

The calculated VPF is >10 times the target cost per life saved criterion of £2M and therefore the proposal should be rejected.

6. DETERMINING SIL TARGETS

Demand Mode and Continuous Mode Safety Functions

When assessing a safety system in terms of fail to function, two main options exist, depending on the mode of operation. If a safety system experiences a low frequency of demands, typically less than once per year, it is said to operate in demand mode. An example of such a safety system is the airbag in a car.

The brakes in a car are an example of a safety system with a continuous mode of operation: they are used (almost) continuously. For demand mode safety systems it is common to calculate the average probability of failure on demand (PFD), whereas the probability of a dangerous failure per hour (PFH) is used for safety systems operating in continuous mode.

Demand Mode Safety Function

For example, assume that in our factory we have on average 1 fire every 2 years and if we did nothing else, this fire would result in a fatality. We could draw a graph of our fatality frequency, Figure 17. The fatality frequency is 0.5 /yr.

FIGURE 17. FATALITY FREQUENCY

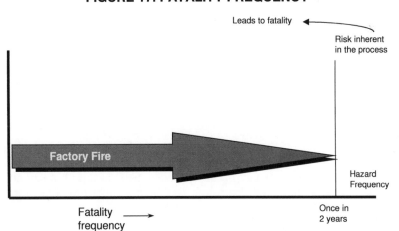

It is vital in this case, that when working out the consequences of the fire, we do not take into consideration any existing safety measures that may be in place. We are looking for the worst case consequences.

If we then fitted a smoke alarm which, let us say, worked 9 times out of 10, then we would expect a fatality on the one occasion, in 10 fires, that the smoke alarm failed to operate on demand. In this case our fatality frequency would decrease from 1 fatality in 2 years to 1 fatality in 20 years.

FIGURE 18. REDUCED FATALITY FREQUENCY

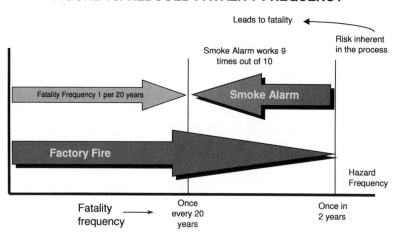

In this example, if the smoke alarm works for 9 fires out of 10, then it has a Probability of Failure on Demand (PFD) of 1 in 10, or 10%. In this case, PFD = 0.1. The smoke alarm with a PFD of 0.1 would reduce the fatality frequency by a factor of 10, giving a Risk Reduction Factor (RRF) of 10.

Summarising, PFD = 1 / RRF.

It is useful to remember that mathematically, PFD is a probability and therefore has a value between zero and 1. It is also dimensionless, which means that simple dimensional checks can be carried out on any SIL target calculations we do.

Example Safety Integrity Level Target

The recommended approach to determining a SIL target is to calculate the risk reduction required to bring the frequency of the consequences of a hazard down to a tolerable level. If we conduct a HAZOP on our plant and identify a hazard in the process that has the potential to do harm, then we must evaluate the potential consequences. The worst case consequences will determine the maximum tolerable frequency for that hazard.

For example, if our hazard could lead to an employee fatality then, based on the tolerability and acceptability of risk criteria, we may allocate a maximum tolerable frequency for the hazard. In other words for the identified hazard of employee fatality, we can specify a maximum tolerable risk of 10^{-4} per year.

By analysing the initiating causes of the hazard, we may estimate the likelihood of the hazard, assuming we do nothing else, and compare it to the specified maximum tolerable frequency. We may do some analysis for example, and determine that our hazard, if unchecked, could occur once per year. This therefore presents a risk gap: something that must be addressed, Figure 19.

FIGURE 19. RISK GAP

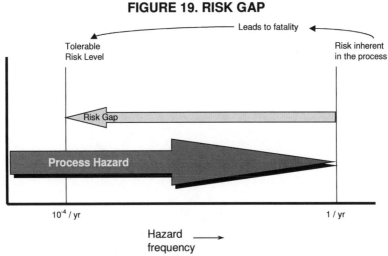

We can then take credit for any safeguards that may already exist to reduce the hazard frequency, such as an alarm, Figure 20. In this case, the alarm reduces the frequency of the hazard consequence by its PFD.

FIGURE 20. TAKING CREDIT FOR ALARMS

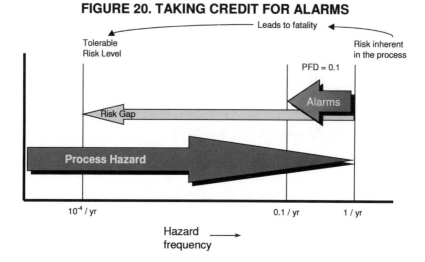

Therefore the risk gap is reduced but the overall residual risk, though smaller, is still greater than the maximum tolerable risk.

Taking other layers of protection into account can reduce the residual risk further. There may be mechanical devices such as a pressure relief valve, a blast wall or bund. Other risk reduction measures may include process control, instrumentation or operator action and each may reduce the residual risk, Figure 21 by their respective PFDs.

FIGURE 21. OTHER LAYERS OF PROTECTION

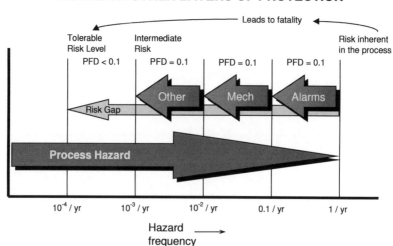

In this example, we have taken all allowed credit for the various safeguards that exist on the plant and we still have a residual risk gap. We can see that to reduce the hazard frequency to less than the maximum tolerable frequency we require another layer, with a PFD of less than 0.1. This is the task of the SIS, Figure 22.

FIGURE 22. PFD TARGET

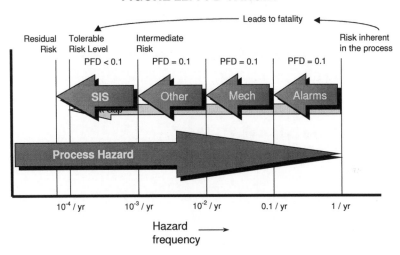

This calculation, albeit carried out graphically here, provides the target PFD for our SIS and enables the SIL target to be determined. This is an example of a Demand Mode Safety Function.

Safety Functions

Typically, the arrangement of a SIS and process is shown, Figure 23.

FIGURE 23. SAFETY INSTRUMENTED SYSTEM

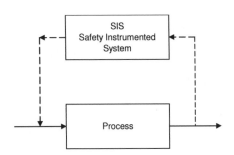

The safety instrumented function monitors some process parameter and takes executive action to make the process safe, if certain limits are passed. This is an example of a demand mode safety function.

A simple example is shown in Figure 24. In the arrangement, the process and SIS are highlighted

FIGURE 24. EXAMPLE DEMAND MODE SAFETY FUNCTION

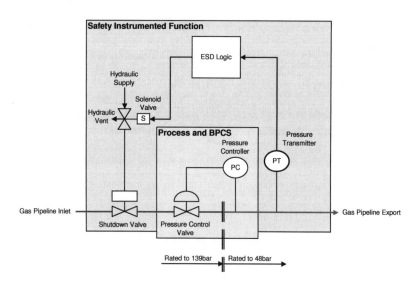

The figure shows a gas pipeline which provides a feed to a power station. The gas passes from left to right, through a shutdown valve, before it reaches the Pressure Control Valve (PCV). The PCV is controlled by a Pressure Controller (PC) which maintains the pressure of the gas to below 48barg, the safe rating of the export pipeline. Failure of this pressure control function could lead to over pressurisation of the downstream pipeline, possible rupture, ignition and fatality and so a safety function has been engineered to safeguard against this scenario.

The safety function consists of a separate Pressure Transmitter (PT), some Emergency Shutdown (ESD) Logic and a Shutdown Valve (SDV) which is actuated by a hydraulic Solenoid Operated Valve (SOV) to shut off the gas supply in the event of the downstream pressure exceeding a pre-set trip level.

The key characteristics of a demand mode safety function are:

- it is generally separate from the process;

- failure of the safety function results in loss of protection, but is not in itself hazardous;

- the frequency of demands placed upon it are low, less than once per year.

Demand mode safety functions include Process Shutdown (PSD), Emergency Shutdown (ESD), and High Integrity Pipeline Protection Systems (HIPPS).

It is often a point of confusion that the PT that forms part of the safety function is providing continuous monitoring of process pressure but that does not preclude it from being demand mode. The term demand mode relates to the frequency of demands for executive active, e.g. the frequency of high pressure excursions.

Demand Mode SIL Targets

IEC61511-1, 9.2.4 groups PFD targets into bands, or Safety Integrity levels (SILs). In the above example, we have a PFD target of <10-1 for our safety function, and this gives a SIL1 requirement as shown, Table 3.

TABLE 3. DEMAND MODE SIL TARGETS

Demand Mode of Operation (Average probability of failure to perform its design function on demand)	Safety Integrity Level
$\geq 10^{-5}$ to $< 10^{-4}$	4
$\geq 10^{-4}$ to $< 10^{-3}$	3
$\geq 10^{-3}$ to $< 10^{-2}$	2
$\geq 10^{-2}$ to $< 10^{-1}$	1

Note: the PFD target is grouped into SIL bands because the standard requires an appropriate degree of rigour in the techniques and measures that are applied in the control and avoidance of systematic failures.

Example of a Continuous Mode Safety Function

An example of a continuous mode safety function is a Burner Management System (BMS) which is used to control a furnace. The system controls the fuel gas and combustion air into the furnace and monitors the burner flame with flame detectors.

Figure 25 shows a typical arrangement of a BMS continuous mode safety instrumented function.

FIGURE 25. CONTINUOUS MODE SAFETY FUNCTION

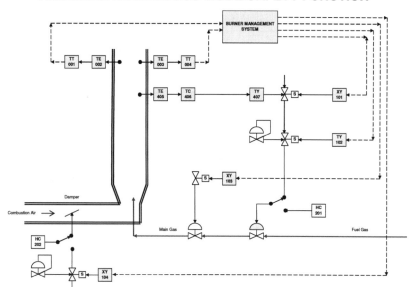

On conditions of flame out, the BMS must shut off the fuel gas to prevent a build-up and possible explosion. Similarly, before lighting, the burner must be purged to ensure that gas has not accumulated within the furnace due to seepage past the valves, or control failures.

So the BMS must provide control through the start-up sequence, purging adequately and it must also monitor operation after lighting. In this example, the BMS and all the associated sensors and valving, constitute a continuous mode safety function.

The key characteristics of a continuous mode safety function are:

- it generally provides some control function;

- failure of the safety function usually leads to a hazardous situation;

- the frequency of demands placed upon it are high, more than once per year or even continuous.

Continuous mode safety functions typically include Burner Management and Turbine Control Systems.

Continuous Mode SIL Targets

IEC61511-1, 9.2.4 also provides SIL Targets for continuous mode systems, Table 4.

TABLE 4. CONTINUOUS MODE SIL TARGETS

Continuous Mode of Operation (Probability of dangerous failure per hour, PFH)	Safety Integrity Level
$\geq 10^{-9}$ to $<10^{-8}$	4
$\geq 10^{-8}$ to $<10^{-7}$	3
$\geq 10^{-7}$ to $<10^{-6}$	2
$\geq 10^{-6}$ to $<10^{-5}$	1

Note: the target failure measure for continuous mode targets is a failure PFH or failure rate.

Footnote.

At first glance, these failure rate targets may seem more onerous than the targets for demand mode systems, e.g. SIL1 (demand mode) should have a PFD of $<10^{-1}$ whereas SIL1 (continuous mode) has a PFH of $<10^{-5}$ failures/hour.

The tables can be aligned however, if we convert the continuous mode targets from failures/hour into failures/year. There are approximately 10^{-4} hours in a year (actually 8760) and so the continuous mode table can be amended as shown in Table 5.

TABLE 5. CONTINUOUS MODE SIL TARGETS (PA)

Continuous Mode of Operation (Probability of dangerous failure per year)	Safety Integrity Level
$\geq 10^{-5}$ to $< 10^{-4}$	4
$\geq 10^{-4}$ to $< 10^{-3}$	3
$\geq 10^{-3}$ to $< 10^{-2}$	2
$\geq 10^{-2}$ to $< 10^{-1}$	1

Modes of Operation (Demand and Continuous Mode Systems)

In determining the way in which a SIS operates, IEC61511-1, 3.2.43 offers the following definitions.

Demand Mode: where a specified action is taken in response to process conditions or other demands. In the event of a dangerous failure of the SIF a potential hazard only occurs in the event of a failure of the process of BPCS.

Continuous Mode: where in the event of a dangerous failure of the SIF a potential hazard will occur without further failure unless action is taken to prevent it.

A good rule of thumb in deciding whether your safety function is continuous or demand mode, is to identify the meaningful metric, or reliability measure.

For example, the air-bags on a car provide a very valuable safety function and as a driver, I would be interested in their probability of failure on demand which indicates that it is a demand mode function. The key characteristics of a demand mode safety function are:

- it is generally separate from the process;
- failure of the safety function results in loss of protection, but is not in itself hazardous.

Table 3 therefore confirms that the targets for demand mode functions are probability of failure on demand.

Alternatively, for the brakes in a car, the meaningful metric would be a failure rate, or a probability of failure per hour. As the driver, I would be very interested in the failure rate of the safety function, so this is a good indication that it is a continuous mode function.

In support of this, the key characteristics of a continuous mode safety function are:

- it generally provides some control function, in this case braking;
- failure of the safety function usually leads to a hazardous situation, loss of speed control.

Table 4 confirms that the continuous mode targets are probability of dangerous failure per hour.

Demand Mode Safety Function Example

Question: A process area is manned for 2 hours per day. Overpressure of the process will result in a gas leak and it is estimated that 1 in 10 gas leaks will cause an explosion that will result in the death of the operator.

Analysis indicates that the overpressure condition will occur every 5 yrs (a rate of 0.2 pa). Assume that the maximum tolerable frequency for the hazard (operator killed by explosion) is 10^{-4} pa.

What is the required PFD of the SIS?

Answer: The fatality rate is:

$$= 0.2 \text{ pa} \times 2/24 \times 1/10$$

$$= 1.67 \times 10^{-3} \text{ pa}$$

Therefore the safety system must have a PFD of:

$$= 10^{-4} \text{ pa} / 1.67 \times 10^{-3} \text{ pa}$$

$$= 6.0 \times 10^{-2}, \text{ which is equivalent to SIL1.}$$

This is an example of a demand mode SIS in that it is only called upon to operate at a frequency determined by the failure rate of the equipment under control.

We can confirm that the result is actually a PFD because we have divided a rate by a rate to give a dimensionless quantity, i.e. a probability.

Continuous Mode Safety Function Example

Figure 26 presents a simple example of a continuous mode safety function. The chemical in the boiler is heated by an electric element which is controlled by a temperature transmitter measuring the outlet.

FIGURE 26. CONTINUOUS MODE SAFETY FUNCTION

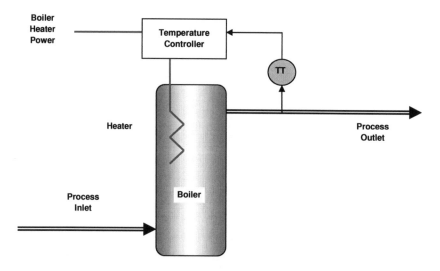

Assume that overheating of the boiler leads to rupture, chemical release and subsequent fire, with the potential for a fatality. There is clearly a risk that should be managed. In this example, the failure rate of the entire process should not exceed the maximum tolerable risk for the hazard.

Question: Assume failure of the boiler leads to overheating and fire and that 1 in 400 failures leads to a fatality. Assume also that the maximum tolerable fatality rate is 10^{-5} pa (third party fatality).

What is the maximum tolerable failure rate of the boiler?

Answer: Since 1 in 400 failures must be less than or equal to the maximum tolerable risk, we can say:

$$10^{-5} \text{ pa} \geq \lambda_B \times 1/400$$

Where λ_B is the failure rate of the boiler.

Therefore:

$$\lambda_B = 400 \times 10^{-5} \text{ pa}$$

$$= 4.0 \times 10^{-3} \text{ pa, which is equivalent to SIL2.}$$

This is an example of a continuous mode SIS that is continually at risk, i.e. in continuous use. The boiler system is allowed to fail 400 times more frequently than the maximum tolerable failure rate because only 1 in 400 failures results in the fatality.

In this example, we would have to design and build the process, i.e. the boiler, heating element and temperature sensor to SIL2 and the failure rate would have to be less than 4.0×10^{-3} pa. This would be a challenging project, but there is another way.

Assume that we have built our boiler process and calculated its failure rate to be 5.0×10^{-2} pa which far exceeds the target of 4.0×10^{-3} pa.

If this was the case, and 1 in 400 failures leads to a fatality, then the fatality frequency would be:

$$= 5.0 \times 10^{-2} \text{ pa} \times 1/400$$

$$= 1.25 \times 10^{-4} \text{ pa}$$

which exceeds the maximum tolerable rate of 10^{-5} pa (third party fatality).

An alternative approach might be to allow the boiler to fail at this unsatisfactorily high rate and engineer a demand mode safety function to bring the fatality frequency down to the maximum tolerable rate, Figure 27.

FIGURE 27. DEMAND MODE SAFETY FUNCTION

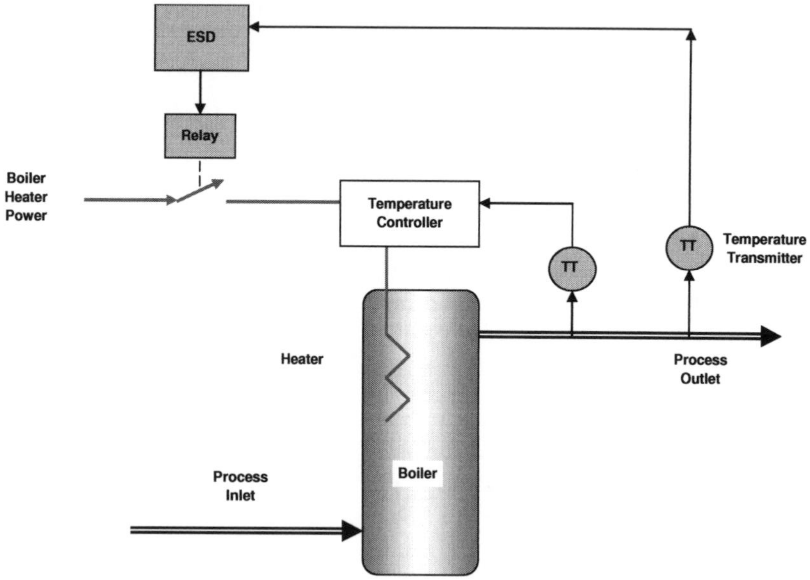

In this configuration, we have a second independent temperature transmitter measuring the outlet temperature and tripping power to the electric heater via some ESD logic, on failure of the process.

We can say:

$$10^{-5}\,pa \geq \lambda_B \times PFD_T$$

where λ_B is the failure rate of the boiler, 1.25×10^{-4} pa and PFD_T is the probability of failure on demand of the independent trip.

Therefore:

$$PFD_T \leq 10^{-5}\,pa\,/\,1.25 \times 10^{-4}\,pa$$

$$PFD_T \leq 0.08$$

which is equivalent to a SIL1 demand mode safety function.

Note: these two examples give us the option of designing the whole boiler system and equipment under control, to SIL2, or we can allow the boiler system to fail and protect it with a SIL1 demand mode safety function. Both options meet the maximum tolerable risk target but engineering a small demand mode SIL1 system is a more cost effective option than a SIL2 continuous boiler control system.

7. RISK GRAPHS

Introduction

Section [6] shows a method of determining SIL targets by calculation but Risk Graphs offer a useful alternative, especially if there are many hazards to analyse. The risk graph method is a useful fast-track technique to apply when there are many hazards to assess.

Figure 28 shows a typical risk graph.

FIGURE 28. TYPICAL RISK GRAPH

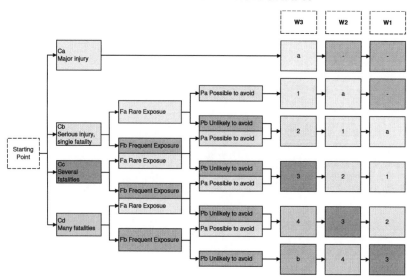

From the starting point, firstly the hazard consequences are determined, Ca, Cb, Cc or Cd.

Then the frequency or exposure of the person most at risk from the hazard, must be estimated and a choice made between Fa, rare exposure, or Fb, frequent exposure. Typically, if the person most at risk has a probability of being within range of the hazardous effects, of 10% or less, then rare exposure can be selected. Otherwise, the exposure can be considered, frequent.

Moving along the risk graph, if the person at risk is likely to be able to avoid the hazard, e.g. by escaping, by being alerted or by being protected by some feature, then we can say it is possible to avoid the hazard and that option can be chosen on the risk graph. Otherwise we must assume that the hazard is unlikely to be avoided and we will arrive at some point, one of the rows in the columns to the right of the risk graph.

Finally, we must select the probability that the hazard will occur by choosing either column W3 (relatively high probability of occurrence), W2 (slight probability of occurrence) or W1 (very slight probability of occurrence). Where the selected row and column meet, then we can read off the required SIL.

Example

As an example, let us assume that a petroleum storage tank could overfill, release vapour which could ignite and lead to several fatalities on site. We have assessed the frequency of filling operations and decided that the probability of the hazard occurring could be W1 (very slight probability). There are no means whereby the plant employees can avoid the hazard should it occur. The plant staff are on site only rarely for maintenance activities, typically for less than 1 hour per day. Figure 29 shows how the risk graph may be used to obtain a SIL target.

FIGURE 29. EXAMPLE USING RISK GRAPH

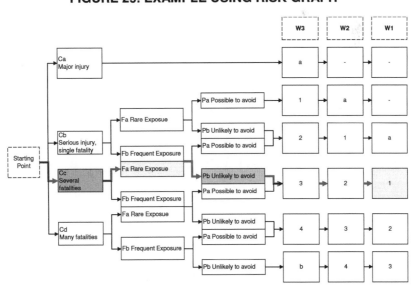

In this example, the safety function could be a high level trip which closes the tank inlet valve. This would have a SIL1 target.

In the risk graph shown, some of the SIL target boxes are labelled 'a' and 'b'. The terms SILa and SILb are sometimes used in industry even though they do not appear in the standard. SILa usually means that some risk reduction should be provided but the risk reduction factor does not need to be as high as SIL1. In other words a PFD of between 1 (no risk reduction) and 0.1, SIL1 is required. Note that some organisations may refer to 'SILa' as '(SIL1)'. SILb is shown in a position higher than SIL4. Generally if you have a SIL4 requirement, it is recommended that the process is reviewed because it is just too dangerous. A SILb requirement is even more dangerous.

Process Industry Risk Graph

Figure 30 shows an example Risk Graph that is similar to those used in the process industry.

FIGURE 30. PROCESS INDUSTRY RISK GRAPH

Consequence Severity	Personnel Exposure	Alternatives To Prevent Danger	High	Medium	Low
Minor Injury			-	-	-
Serious Injury or Single Fatality	Low Exposure	Possible Avoidance	1	-	-
		Avoidance Not	2	1	-
	High Exposure	Possible Avoidance	2	1	1
		Avoidance Not	3	2	1
Multiple Fatalities	Low Exposure		3	3	2
	High Exposure		NR	3	3
Catastrophic			NR	NR	NR

Demand Rate

Requirement Categories

- = No Special safety Features Required
NR = Not Recommended
High = 0.5 to 5 pa
Medium = 0.05 - 0.5 pa
Low < 0.05 pa

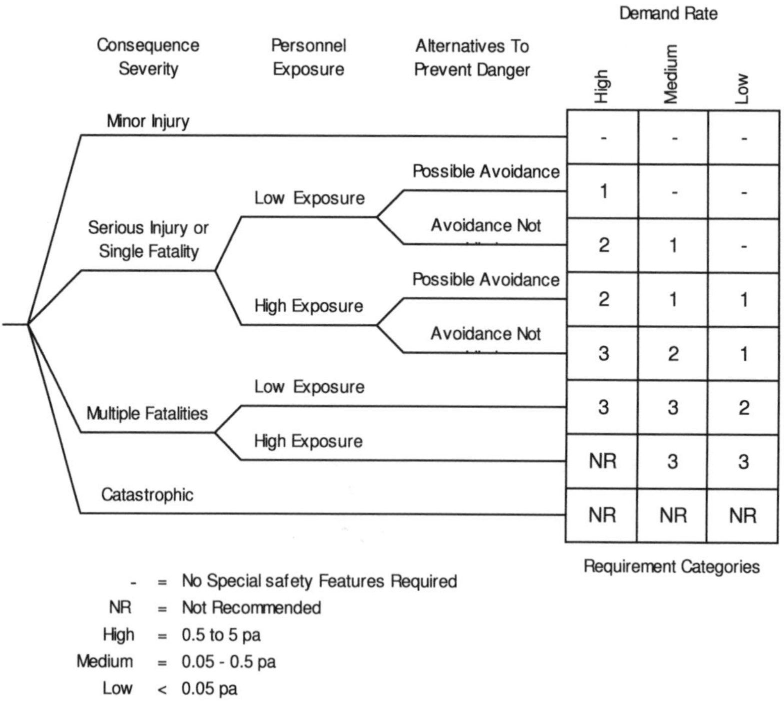

The principle of use is exactly the same as for the risk graph shown in Figure 28 but in this case some guidance is provided on the estimation of demand rate. Risk graphs can be subjective and can suffer from the problem of interpretation of risk parameters. Thus, it can lead to inconsistent outcomes that may result in pessimistic SIL targets.

If for example, a hazard could result in multiple fatalities, with rare exposure and a demand rate of 0.05 /year, then the demand rate falls somewhere between the 'Low' and 'Medium' categories and a decision will have to be made as to which column to choose. Taking a conservative approach would result in a SIL3 target, Figure 31.

FIGURE 31. EXAMPLE USING RISK GRAPH

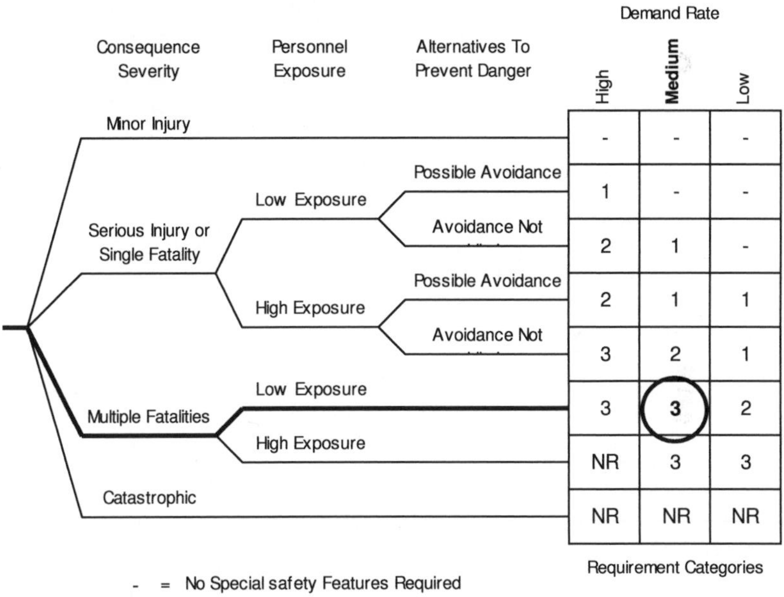

Consequence Severity	Personnel Exposure	Alternatives To Prevent Danger	Demand Rate		
			High	Medium	Low
Minor Injury			-	-	-
Serious Injury or Single Fatality	Low Exposure	Possible Avoidance	1	-	-
		Avoidance Not	2	1	-
	High Exposure	Possible Avoidance	2	1	1
		Avoidance Not	3	2	1
Multiple Fatalities	Low Exposure		3	3	2
	High Exposure		NR	3	3
Catastrophic			NR	NR	NR

Requirement Categories

- = No Special safety Features Required
NR = Not Recommended
High = 0.5 to 5 pa
Medium = 0.05 - 0.5 pa
Low < 0.05 pa

A less cautious interpretation would have resulted in a SIL2 target.

Example

Figure 32 shows an example of a typical Risk Matrix. The P, A, E and R columns provide descriptions of possible hazard consequences, the frequencies of occurrence are described in qualitative terms and target SILs are provided where the rows and columns align.

FIGURE 32. EXAMPLE RISK MATRIX

P People	A Asset	E Environmental	R Reputation	A <0.01/yr Never heard of in industry	B <0.05/yr Has occurred in industry	C <0.25/yr Has occurred in the company	D <2/yr Happens several times /year in the company	E >2/yr Happens several times /year in the facility
No injury	No damage	No effect	No effect					(SIL1)
Slight injury (<1/yr)	Slight damage (<$10k)	Slight effect	Slight impact				(SIL1)	SIL1
Minor injury (<1E-01/yr)	Minor damage (<$100k)	Minor effect	Minor impact			(SIL1)	SIL1	SIL2
Major injury (<1E-02/yr)	Major damage (<$500k)	Localised effect	Considerable impact		(SIL1)	SIL1	SIL2	SIL3
Single fatality (<1E-03/yr)	Major damage (<$10M)	Major effect	National impact	(SIL1)	SIL1	SIL2	SIL3	N/A
Multiple fatalities (<1E-04/yr)	Extensive damage (>$10M)	Massive effect	International impact	SIL1	SIL2	SIL3	N/A	N/A

This looks to be a straightforward and useful approach but there could be potential problems if care is not taken.

A: The frequencies of occurrence must be quantified in a way that is not only consistent with the description, but also this must result in the correct SIL target.

B: "Never heard of in industry" may be estimated by assuming say, 5000 plants operating over 20 years which would give a frequency of say $<10^{-5}$ /year and not $<10^{-2}$ /year as shown. With a maximum tolerable risk of $<10^{-4}$ /year this would result in no SIL target.

C: The maximum tolerable risk frequencies must be appropriate. A value of $<10^{-3}$ /year for a single fatality is too high and will result in optimistic SIL targets and inadequate risk reduction.

D: For the target SILs to increment by row and by column as they do in Figure 32, the frequencies of occurrence would also have to increase by an order of magnitude between each column.

E: The SIL target of (SIL1) means that some risk reduction is required but there is no consequential effect. No protection is required if there is no hazardous event.

F: Finally, for the commercial categories, the frequency of occurrence of asset damage must be realistic and consistent with the cost of implementing the required SIF.

The risk matrix therefore requires calibration and the following is suggested, Figure 35.

FIGURE 33. RISK MATRIX CALIBRATION

P People	A Asset	E Environmental	R Reputation	<1E-04/yr A Never heard of in industry	<1E-03/yr B Has occurred in industry	<1E-02/yr C Has occurred in the company	<0.1/yr D Happens several times /year in the company	>0.1/yr E Happens several times /year in the facility
No injury	No damage	No effect	No effect					
Slight injury (<0.1/yr)	Slight damage (<$10k)	Slight effect	Slight impact				(SIL1)	SIL1
Minor injury (<1E-02/yr)	Minor damage (<$100k)	Minor effect	Minor impact			(SIL1)	SIL1	SIL2
Major injury (<1E-03/yr)	Major damage (<$500k)	Localised effect	Considerable impact		(SIL1)	SIL1	SIL2	SIL3
Single fatality (<1E-04/yr)	Major damage (<$10M)	Major effect	National impact	(SIL1)	SIL1	SIL2	SIL3	N/A
Multiple fatalities (<1E-05/yr)	Extensive damage (>$10M)	Massive effect	International impact	SIL1	SIL2	SIL3	N/A	N/A

Summary

Risk graphs and risk matrices can be very useful, particularly when used as a first pass, fast-track technique to screen out all but the higher SILs, i.e. SIL2 and above. However, a careful calibration of the techniques used should avoid incorrect results as a result of some of the pitfalls shown here.

8.   LAYER OF PROTECTION ANALYSIS (LOPA)

Introduction

Layers of Protection Analysis (LOPA) is a structured way of calculating risk reduction (and SIL) targets.

LOPA is carried out in a similar forum to a HAZOP. Potential hazards are typically identified using the HAZOP approach [3] and imported into the LOPA worksheets, thus maintaining a traceable link between the two analyses from hazard identification, through to risk reduction requirement and SIL target. The LOPA may be carried out as an extension to the HAZOP meeting as there is a natural progression from one to the other.

LOPA Study Team

It is important that a LOPA Study Team is made up of personnel who will bring the best balance of knowledge and experience, of the type of plant being considered, to the study. A typical LOPA team is made up as follows:

Name	Role
Chairman	To explain the LOPA process, to direct discussions and facilitate the LOPA. Someone experienced in LOPA but not directly involved in the design, to ensure that the method is followed carefully.
Secretary	To capture the discussion of the LOPA Meeting and provide on-line analysis of SIL Targets. To log recommendations or actions.
Process Engineer	Usually the chemical engineer responsible for the process flow diagram and development of the Piping and Instrumentation Diagrams (P&IDs).
User / Operator	To advise on the use and operability of the process, and the effect of deviations.
C&I Specialist	Someone with relevant technical knowledge of Control and Instrumentation.
Maintainer	Someone concerned with maintenance of the process.

Name	Role
A design team representative	To advise on any design details or provide further information.

Information Used in the LOPA

The following items should be available to view by the LOPA team:

- P&IDs for the facility;
- Process Description or Philosophy Documents;
- Existing Operating and Maintenance Procedures;
- C&E charts;
- Plant layout drawings.

Establishing SIL Targets

The LOPA technique, as described in AIChE Centre for Chemical Process Safety document, Layer of Protection Analysis, 2001 [20.6] can be used to establish SIL targets.

LOPA considers hazards identified by other means, e.g. HAZOP but LOPA can be conducted as part of a HAZOP Meeting, evaluating each hazard as they are identified.

The LOPA Team considers each hazard identified and documents the initiating causes and the protection layers that prevent or mitigate the hazard. The total amount of risk reduction is then determined and the need for more risk reduction analysed. If additional protection is to be provided in the form of a SIS, the methodology would allow the determination of the appropriate SIL and the required PFD.

The LOPA process is recorded on LOPA Worksheets which allow the initiating events and their frequencies to be quantified together with the risk reduction provided by the independent layers of protection to be claimed. The worksheet headings are described in the following sections and an example of a LOPA is provided.

Example LOPA

Consider the separator vessel example, Figure 10, page 27, the identified hazards can be imported into the LOPA worksheet and the risks analysed. The worksheets are presented on page 82 and the following sections describe the worksheet headers and provide guidance on quantification.

Hazard ID / ref

Provides an identifier for each hazard. In the example, the hazard considered for analysis is ref. **1.10: High Pressure in Vessel**. This reference provides backward traceability with other studies, in this case, the HAZOP and as the project proceeds, will provide forward traceability with SIF Allocation and SIL Verification.

Event (Hazard) Description

Provides a description of the potential hazard identified.

Consequence

Describes the consequence of the hazard. In the example LOPA, we have analysed the consequences of the hazard in terms of personnel safety, risks to the environment and also risks to the asset, i.e. commercial risks.

Severity Category (Sev Cat)

The severity of the documented consequences may be categorised and derived from a Risk Classifications table, for example Table 6.

Maximum Tolerable Risk (MTR)

The maximum tolerable frequency of the hazard consequence as applied to personnel safety, but typically also applied to the environment, to the reputation of the organisation and to potential damage to the environment, to the reputation of the company and the commercial costs resulting from damage to the asset, loss of revenue or security of supply. The maximum tolerable frequencies used should be in line with HSE guidance, e.g. R2P2 [20.5] for safety.

However, the maximum tolerable frequencies for environmental, reputation and commercial risks should be a company decision.

Typical values that could be used are shown in Table 6.

TABLE 6. RISK CRITERIA

Consequence	Sev. Cat.	Risk Target Frequency (/yr)	Consequence Description	
			On Site	Off site
People (Safety)	P1	1.00E-01	Employee medical treatment or Restricted Work Injuries	Medical Treatment or Restricted Work Injuries (3rd party)
	P2	1.00E-02	Employee Lost Time Accident (LTA) with no permanent effect	LTA (3rd party) with no permanent effect
	P3	1.00E-03	Employee permanent effect	No permanent effects
	P4	1.00E-04	1 employee fatality and/or several permanent disabilities	Permanent effects (3rd party)
	P5	1.00E-05	Several employee fatalities (2 - 10)	Single 3rd party fatality and/or many permanent disabilities
	P6	1.00E-06	Many employee fatalities (over 10)	Several 3rd party fatalities
Environment	E1	1.00E-01	No declaration to authorities, but clean-up required	No declaration to authorities, but minor clean-up required. (e.g. spill of 1- 100 litres with kit deployed)
	E2	1.00E-02	Declaration to authority, but no environmental consequences	Declaration to authority, but no environmental consequences. (e.g. spill of > 100 litres in bunded/ interceptored customer premises)
	E3	1.00E-03	Moderate pollution within site limits	Moderate pollution requiring remediation works (e.g. with plume leaving site, but with site remaining operational)

Consequence	Sev. Cat.	Risk Target Frequency (/yr)	Consequence Description	
			On Site	Off site
Environment	E4	1.00E-04	Significant pollution within site limits. Evacuation of persons / temp. Site closure OR Significant pollution external to site. Evacuation of persons. (e.g. off -site spill at service station)	Significant pollution external to site. Evacuation of persons. (e.g. off -site spill at service station)
	E5	1.00E-05	see off site consequences	Important pollution with reversible environmental consequences external to site. (e.g. Major Accident To The Environment)
	E6	1.00E-06	see off site consequences	Major and sustained pollution external to site and/or extensive loss of aquatic life (e.g. loss of ship cargo)
Cost	C1	1.00E-01	<£10K loss	N/A
	C2	1.00E-02	£10K < £100K loss	N/A
	C3	1.00E-03	£100K < £1.0M loss	N/A
	C4	1.00E-04	£1.0M < £10M loss	N/A
	C5	1.00E-05	£10M < £100M loss	N/A
	C6	1.00E-06	$\geq$ £100M loss	N/A
Reputation	R1	1.00E-01	No publicity. Locals affected.	N/A
	R2	1.00E-02	Local press	N/A
	R3	1.00E-03	National press	N/A
	R4	1.00E-04	National television	N/A
	R5	1.00E-05	International press	N/A
	R6	1.00E-06	International television	N/A

Note, when applied to personal safety, this represents the frequency that the individual most at risk, is exposed to the hazard.

Initiating Cause

Lists the identified causes of the hazard. These causes are determined during the LOPA Meeting from the experience of the attendees. For the example hazard, overpressure, the potential initiating causes, their frequencies of occurrence and the data source is presented in Table 7. The LOPA should provide visibility of all data by presenting all initiating events and frequencies, with reference to data sources, in this way.

TABLE 7. INITIATING EVENTS AND FREQUENCIES

Initiating Cause	Initiating Likelihood (/yr)	Data Source
DCS fails to control pressure.	1.65E-02	Exida 2007, item x.x.x
Liquid level LL101 fails and reads low level.	1.10E-02	Exida 2007, item x.x.x
TT100 fails and reads low temperature.	2.68E-03	Exida 2007, item x.x.x
PT102 fails and reads low pressure.	8.58E-04	Exida 2007, item x.x.x
Gas Export FCV102 fails closed.	1.01E-02	Oreda 2009, item x.x.x
Fuel Gas FCV100 fails open.	1.01E-02	Oreda 2009, item x.x.x
Liquid Export XV102 fails closed.	2.89E-03	Oreda 2009, item x.x.x
Liquid Import XV102 fails open.	2.89E-03	Oreda 2009, item x.x.x

Initiating Likelihood (/yr)

Quantifies the expected rate of occurrence of the initiating cause. This rate can be estimated based on the experience of the attendees and any historical information available or it may be derived from suitable failure rate sources.

The initiating events and their frequencies of occurrence for the example are presented in Table 7. Where initiating likelihoods are based on human factors such as operator error, this can be challenging to estimate. One technique is to base the estimation on the frequency of opportunities that an operator has to make an error, and factor this by the probability that they will make a dangerous error.

For example, let us assume an operator can initiate an overpressure in a pipeline by closing a valve. Normally the operator opens a by-pass valve before closing the main valve and he does this every month. The base frequency λ_B, for this activity, is therefore 12 per year (once per month).

We can assume the operator is well trained, the task is routine and he is not under any stress so we may estimate the probability he will make an error, P_E, e.g. fails to open the by-pass valve first, will be say 1%. The initiating event frequency, λ_{INIT} can be estimated as:

$$\lambda_{INIT} \quad = \quad \lambda_B \times P_E$$

$$\lambda_{INIT} \quad = \quad 12 \times 1\% \text{ /year}$$

$$\lambda_{INIT} \quad = \quad 0.12 \text{ /year}$$

Usually, we can perform a sensibility check on this data by asking the LOPA participants if they have any experience of such an event occurring, or whether they feel the frequency is reasonable. A frequency of 0.12 /year is equivalent to an error every 8 years.

Conditional Modifiers

Leak Size Distribution, column [b]

In the example, the postulated consequences of the overpressure hazard will only occur if the pressure condition results in a vessel rupture. Most overpressure conditions, it could be argued, would result in no loss of containment, or a minor leak from a flange for example. In the example, the LOPA Team estimated that 10% of the initiating events would result in the consequences.

Probability of Ignition, column [c]

For the postulated safety and commercial consequences, we require the released gas to ignite. In this example, we have made reference to a fire safety study which predicted a 75% probability of ignition, given a large rupture scenario. For the safety consequences therefore, we can claim 0.75 as a conditional modifier and the initiating event frequency will be reduced by this factor.

For the environmental consequences, no risk reduction can be claimed since ignition is not necessary for the consequences.

General Purpose Design, column [d]

An example of general purpose design would be a jacketed pipe which would provide some protection from a loss of containment. In the example, no credit was taken for general purpose design because there were no specific design features that provided any risk reduction.

Independent Protection Layers (IPLs)

Each protection layer consists of a grouping of equipment and/or administrative controls that function in concert with the other layers.

The level of protection provided by each IPL is quantified by the probability that it will fail to perform its specified function on demand, its PFD, a dimensionless number between 0 and 1. The smaller the value of the PFD, the larger the risk reduction factor that is applied as a modifying factor, to the calculated initiating likelihood, hence where no IPL is claimed, a '1' is inserted into the LOPA worksheet.

In the example, the IPLs claimed in columns [e] to [h] can be tailored to suit the application. Typical IPLs have been presented.

Basic Process Control System (BPCS), column [e].

Credit can be claimed if a control loop in the BPCS (DCS) prevents the hazard from occurring as a result of a potential initiating cause. In the example, Figure 10, page 27, for some of the initiating causes, e.g. liquid import valve XV102 fails open, the BPCS (DCS) can compensate for this by opening the liquid export valve and preventing a high level. A PFD of 0.1 has been claimed which means that the DCS will prevent the consequences from occurring in 9 out of 10 events.

A PFD of 0.1 is generally the most risk reduction that can be claimed for a non-SIL rated system. This is because the DCS can be manually adjusted, there is generally not such strict control over trip point settings and the testing regime is not as rigorous as for a SIS.

Independent Alarms, column [f].

Credit can be claimed for alarms that are independent of the BPCS, and alert the operator and utilise operator action. Credit can only be claimed if the alarm is truly independent from the BPCS and the SIF and only if the operator can respond to the alarm and take action to make the process safe, within the safe process time. Typically, a PFD of 0.1 may be claimed for independent alarms. In this example, no credit has been claimed.

Additional Mitigation

Occupancy, column [g].

Access - Mitigation layers can include the occupancy, i.e. the proportion of time that an operator is exposed to a hazard and restricted access to hazardous areas. In this example, an occupancy based on an 8 hour shift has been claimed.

Other Mitigation: column [h].

Additional mitigation may be available in the form of:

- Physical - physical barriers to mitigate the hazard once it has been initiated. Examples would be pressure relief devices and bunds.

- Operator Action – credit can be claimed for detection and inspection at regular intervals provided the operator can take appropriate action.

In the example, no credit has been claimed.

LOPA Calculated Results

Intermediate Level Event Likelihood

The intermediate event likelihood is calculated by multiplying the initiating likelihood by the PFDs of the mitigating layers of protection. The calculated number is in units of events per year. The total intermediate level likelihood indicates the demand rate on any proposed SIF.

SIS Required PFD

Calculated by comparing the Maximum Tolerable Risk λ_{MTR}, with the Intermediate Level Event Likelihood, or the hazard frequency, λ_{HAZ}.

$$\textbf{PFD} \quad = \quad \lambda_{MTR} / \lambda_{HAZ}$$

SIS Required SIL

Obtained from Table 8, corresponding to the SIS Required PFD.

TABLE 8. SIL SPECIFIED PFD AND FAILURE RATES

SIL	DEMAND MODE Probability of failure on demand	CONTINUOUS MODE Failure rate per hour
SIL 4	$\geq 10^{-5}$ to $< 10^{-4}$	$\geq 10^{-9}$ to $< 10^{-8}$
SIL 3	$\geq 10^{-4}$ to $< 10^{-3}$	$\geq 10^{-8}$ to $< 10^{-7}$
SIL 2	$\geq 10^{-3}$ to $< 10^{-2}$	$\geq 10^{-7}$ to $< 10^{-6}$
SIL 1	$\geq 10^{-2}$ to $< 10^{-1}$	$\geq 10^{-6}$ to $< 10^{-5}$

It should be noted that the PFD and failure rate for each SIL, depend upon the mode of operation in which a SIS is intended to be used, with respect to the frequency of demands made upon it.

The following LOPA Worksheets are to be read across double pages.

Example LOPA

ID / Ref.	Zone Description	Event (Hazard) Description	Consequence	Sev Cat	Max Tolerable Risk (pa)	Initiating Cause	Initiating Likelihood (pa)	Leak - size distribution	Prob of Ignition
							[a]	[b]	[c]
						DCS fails to control pressure.	1.65E-02	0.10	0.75
1.10	Vessel	High pressure causes vessel rupture and gas release.	**Safety:** Gas release ignites on burner and hot surfaces. Possibly two maintainer fatalities.	P5	1.00E-05	PT102 fails and reads low pressure.	8.58E-04	0.10	0.75
						Liquid Import XV102 fails open.	2.89E-03	0.10	0.75
						Gas Export FCV102 fails closed.	1.01E-02	0.10	0.75
						Liquid Export XV102 fails closed.	2.89E-03	0.10	0.75
						TT100 fails and reads low temperature.	2.68E-03	0.10	0.75
						Fuel Gas FCV100 fails open.	1.01E-02	0.10	0.75
						Liquid level LL101 fails and reads low level.	1.10E-02	0.10	0.75
						DCS fails to control pressure.	1.65E-02	0.10	
1.10	Vessel	High pressure causes vessel rupture and gas release.	**Environmental:** Vessel rupture, gas escape, no ignition. Release on site. Clean-up and declaration to authority required, but no environmental consequences.	E2	1.00E-02	PT102 fails and reads low pressure.	8.58E-04	0.10	
						Liquid import XV102 fails open.	2.89E-03	0.10	
						Gas Export FCV102 fails closed.	1.01E-02	0.10	
						Liquid Export XV102 fails closed.	2.89E-03	0.10	
						TT100 fails and reads low temperature.	2.68E-03	0.10	
						Fuel Gas FCV100 fails open.	1.01E-02	0.10	
						Liquid level LL101 fails and reads low level.	1.10E-02	0.10	

Example LOPA (continued)

General Purpose Design (Design Rating)	Independent Layers of Protection				Intermediate Level Event Likelihood (pa)	SRS required PFD	SRS required SIL	Comments / Assumptions
	BPCS [DCS]	Indep't Alarms	Additional Mitigation: Occupancy (Manning Levels)	Additional Mitigation e.g. Fire walls / Relief Valves				
[d]	[e]	[f]	[g]	[h]				
			0.33		4.13E-04			[a] Refer to initiating event data. [b] LOPA Team estimate probability of large leak (Rupture) at 10%. [c] Fire risk study estimates probability of ignition 75%. [d] No credit claimed for design features. [e] DCS is the initiating cause, therefore no credit calimed for DCS. [f] No independent alarms available. No credit claimed. [g] Vessel area occupied 8hrs per day. [h] No pressure relief valves. No credit claimed.
			0.33		2.15E-05	1.87E-02	SIL1	As above.
	0.10		0.33		7.23E-06			As above except: [e] DCS can compensate for import valve failures. Estimate PFD = 0.1.
	0.10		0.33		2.52E-05			As above.
	0.10		0.33		7.23E-06			As above.
	0.10		0.33		6.70E-06			As above.
	0.10		0.33		2.52E-05			As above.
	0.10		0.33		2.74E-05			As above.
					5.34E-04			
					1.65E-03			[a] Refer to initiating event data. [b] LOPA Team estimate probability of large leak (Rupture) at 10%. [c] Ignition not required. No risk reduction claimed. [d] No credit claimed for design features. [e] DCS is the initiating cause, therefore no credit calimed for DCS. [f] No independent alarms available. No credit claimed. [g] Environment at risk 24hrs/day. No ridk reduction claimed. [h] No pressure relief valves. No credit claimed.
					8.58E-05	None	None	As above.
	0.10				2.89E-05			As above except: [e] DCS can compensate for import valve failures. Estimate PFD = 0.1.
	0.10				1.01E-04			As above.
	0.10				2.89E-05			As above.
	0.10				2.68E-05			As above.
	0.10				1.01E-04			As above.
	0.10				1.10E-04			As above.
					2.14E-03			

Example LOPA (continued)

ID / Ref.	Zone Description	Event (Hazard) Description	Consequence	Sev Cat	Max Tolerable Risk (pa)	Initiating Cause	Initiating Likelihood (pa)	Leak - size distribution	Prob of Ignition
							[a]	[b]	[c]
1.10	Vessel	High pressure causes vessel rupture and gas release.	**Commercial:** Vessel rupture, gas escape, ignition and damage to asset. Equipment damage requiring vessel replacement estimated at £10M and loss of production for 1 year.	C5	1.00E-05	DCS fails to control pressure.	1.65E-02	0.10	0.75
						PT102 fails and reads low pressure.	8.58E-04	0.10	0.75
						Liquid import XV102 fails open.	2.89E-03	0.10	0.75
						Gas Export FCV102 fails closed.	1.01E-02	0.10	0.75
						Liquid Export XV102 fails closed.	2.89E-03	0.10	0.75
						TT100 fails and reads low temperature.	2.68E-03	0.10	0.75
						Fuel Gas FCV100 fails open.	1.01E-02	0.10	0.75
						Liquid level LL101 fails and reads low level.	1.10E-02	0.10	0.75

Example LOPA (continued)

General Purpose Design (Design Rating)	Independent Layers of Protection				Intermediate Level Event Likelihood (pa)	SRS required PFD	SRS required SIL	Comments / Assumptions
	BPCS [DCS]	Indep't Alarms	Additional Mitigation: Occupancy (Manning Levels)	Additional Mitigation e.g. Fire walls / Relief Valves				
[d]	[e]	[f]	[g]	[h]				
					1.24E-03			[a] Refer to initiating event data. [b] LOPA Team estimate probability of large leak (Rupture) at 10%. [c] Fire risk study estimates probability of ignition 75%. [d] No credit claimed for design features. [e] DCS is the initiating cause, therefore no credit calimed for DCS. [f] No independent alarms available. No credit claimed. [g] Vessel area occupied 8hrs per day. [h] No pressure relief valves. No credit claimed.
					6.44E-05	6.24E-03	SIL2	As above.
	0.10				2.17E-05			As above except: [e] DCS can compensate for import valve failures. Estimate PFD = 0.1.
	0.10				7.56E-05			As above.
	0.10				2.17E-05			As above.
	0.10				2.01E-05			As above.
	0.10				7.56E-05			As above.
	0.10				8.21E-05			As above.
					1.60E-03			

LOPA Results

The results, Table 9 show that the overpressure hazard has safety consequences which may be protected with a SIL1 SIF with a PFD of $\leq$ 1.87E-02. However, the commercial risk dominates and requires a SIL2 SIF with a PFD of $\leq$ 8.24E-03.

TABLE 9. LOPA RESULTS

Hazard	Consequence	SIL Target	PFD Target
Safety	**Safety:** Gas release ignites on burner and hot surfaces. Possibly two maintainer fatalities.	SIL1	1.87E-02
Environmental	**Environmental:** Vessel rupture, gas escape, no ignition. Release on site. Clean-up and declaration to authority required, but no environmental consequences.	None	None
Commercial	**Commercial:** Vessel rupture, gas escape, ignition and damage to asset. Equipment damage requiring vessel replacement estimated at £10M and loss of production for 1 year.	SIL2	6.24E-03

It is not uncommon for non-safety hazards to dominate. In this example, the asset is always at risk from the hazard whereas, in terms of safety, personnel are only at risk part of the time.

The SIF that is to be engineered to protect against overpressure should therefore meet the commercial targets and the same SIF will therefore provide adequate protection to personnel.

9.   ALLOCATION OF SAFETY FUNCTIONS

Lifecycle Phases

Figure 34 shows the phase of the lifecycle that applies.

FIGURE 34. LIFECYCLE PHASE 2

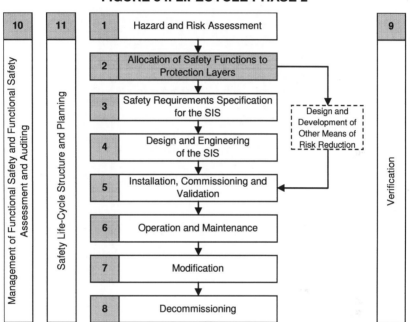

The objective of Lifecycle Phase 2 as defined in IEC61511-1, 9.1 is to allocate safety functions to protection layers.

As inputs, the phase requires a description in terms of the safety function requirements and the safety integrity requirements.

As outputs, the phase is required to provide information on the allocation of the overall safety functions, their target failure measures, and associated safety integrity levels. Assumptions made concerning other risk reduction measures that need to be managed throughout the life of the process or plant should also be defined.

The purpose of safety function allocation is therefore to allocate safety functions to specific protection layers in order to prevent, control or mitigate hazards from the process, Figure 35.

FIGURE 35. SAFETY FUNCTION ALLOCATION

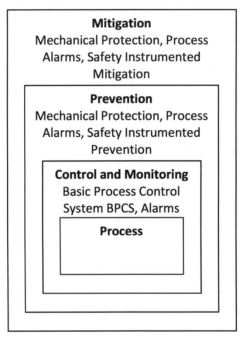

The requirement is to agree on the safety layers to be used and to allocate performance targets for any SIFs that are required.

The preferred approach would be to aim to achieve an inherently safe design. In practice, safety instrumented functions are typically used where there are problems in achieving this. For example a SIF may be used to protect against limitations on relief capacity or to provide protection against exothermic reactions in a vessel.

Decisions on the allocation of safety functions to safety layers are often taken on the basis of what has been found to be good practice by the user organisation, or the industry.

The hazard and risk assessment and the allocation process should provide a clear description of the functions to be carried out by the safety systems, including potential SIFs and their SIL requirements.

This forms the basis of the Safety Requirements Specification (SRS), [10].

At this stage of the implementation, it is unnecessary for the allocation of safety functions, to specify details of technology, or architectural requirements for sensors and valves. Such decisions are complex and depend on many factors. Whether a particular system requires 2oo3 sensors and 1oo2 valves is examined in the Design and Engineering of the SIS [15].

However, in safety function allocation, early consideration should be given to common cause failures between redundant parts within a layer. For example, common cause failures may exist between redundant pressure relief valves on the same vessel, between different safety layers, or between safety layers and the BPCS.

Any sensors or actuators which are shared by the BPCS and SIS are also likely to be at risk from common cause failures. A failure of a BPCS measurement could result in a dangerous failure of the SIF, if a device with the same characteristics is used within the safety instrumented system. In such cases it will be necessary to establish if there are credible failure modes that could cause failure of both devices at the same time.

Common cause failures can be reduced by changing the design of the SIF or the BPCS. The preferred approach is the introduction of diversity in design or technology, and physical separation. These are two effective methods of reducing the likelihood of common cause failures.

The likelihood of the common cause event should also be taken into account when determining whether the overall risk reduction is adequate. This may require fault tree analysis.

Safety Function Allocation

The Separator Vessel example [Figure 10, page 27], is reproduced here to enable the phase of the life-cycle to be considered.

The vessel takes in the process liquid which is heated by a gas burner. Vapour is separated from the process liquid and released for export. The remaining concentrated liquid is drawn off from the bottom of the vessel when the reaction is complete, Figure 36.

The vessel has a Distributed Control System (DCS) which controls liquid level within the vessel, gas pressure and temperature.

FIGURE 36. SEPARATOR VESSEL

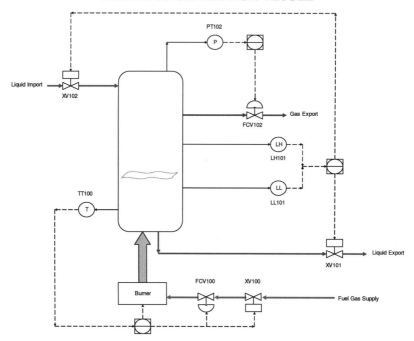

In the example, the following SIFs and SIL requirements were identified, Table 10. SIF Requirements Table 10.

The analysis of Hazard Reference 1.10 was shown as part of the LOPA example. LOPA was also used to determine SIL targets and PFD targets for the other identified hazards.

TABLE 10. SIF REQUIREMENTS

HAZOP Ref	Hazard	Consequence	SIL Target	PFD Target
1.10	High pressure causes vessel rupture and gas release.	Gas release ignites on burner and hot surfaces. Possibly two maintainer fatalities. Equipment damage requiring vessel replacement estimated at £10M and process shutdown for 1 year. Minor environmental release.	SIL2	6.24E-03
1.11	Low pressure causes vessel rupture and gas release.	Gas release ignites on burner and hot surfaces. Possibly two maintainer fatalities. Equipment damage requiring vessel replacement estimated at £10M and process shutdown for 1 year. Minor environmental release.	None	None
1.15	High temperature leads to high pressure, vessel rupture and gas release.	Gas release ignites on burner and hot surfaces. Possibly two maintainer fatalities. Equipment damage requiring vessel replacement estimated at £10M and process shutdown for 1 year. Minor environmental release.	None	None
1.16	Low temperature, potential liquid freezing (solidifying), vessel rupture and loss of containment.	Equipment damage requiring vessel replacement estimated at £10M and process shutdown for 6 months. Environmental release requiring notification order.	None	None
1.20	High level in vessel could result in liquid carry over into gas export.	Equipment damage downstream requiring vessel replacement estimated at £10M and process shutdown for 6 months.	SIL1	8.10E-02
1.21	Low level in vessel could result in gas blow-by into liquid export.	Equipment damage downstream requiring vessel cleaning estimated at £2M and process shutdown for 6 weeks.	SIL1	6.22E-02

The intermediate event likelihood indicated by the LOPA determined that all proposed SIFs would be demand mode functions.

SIL1 targets were established for high level and low level and the following SIFs were therefore proposed.

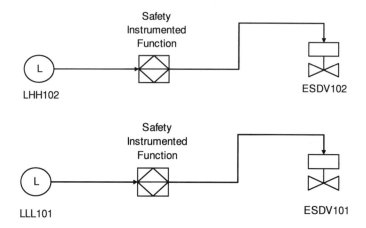

To protect against high pressure, a pressure relief valve was implemented as good engineering practice and an additional SIF established as shown below.

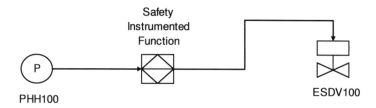

The individual SIFs together form the overall SIS:

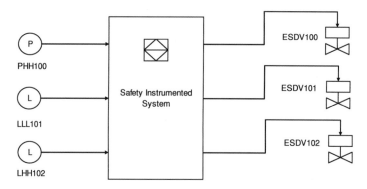

Figure 37 highlights the allocated SIFs:

FIGURE 37. SAFETY FUNCTIONS

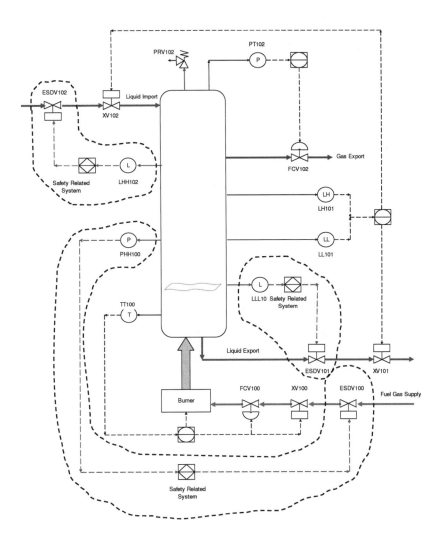

10. SAFETY REQUIREMENTS SPECIFICATION

Lifecycle Phases

Figure 38 shows the phase of the lifecycle that applies.

FIGURE 38. LIFECYCLE PHASE 3

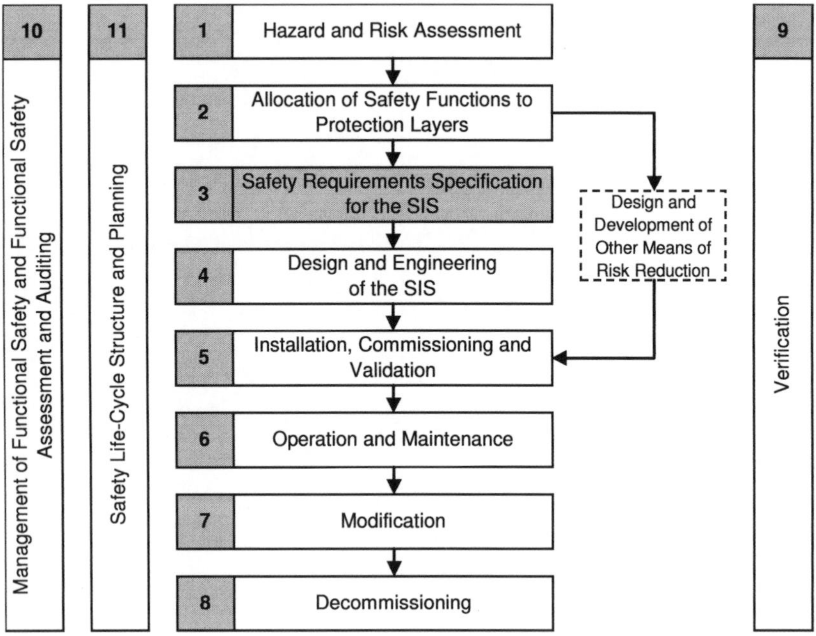

The objective of Lifecycle Phase 3 as defined in IEC61511-1, 10.1 is to specify the requirements for the SIFs.

Safety Integrity Requirements of a SIF

The SIL of each SIF has been selected during the SIL determination study using Risk Graph, LOPA or Risk Matrix.

This information must now be communicated to the design team by the Safety Requirements Specification (SRS) to ensure the design meets the SIF safety integrity requirements during implementation. The SRS is the basis of the SIF validation.

Functional requirements are derived from the hazard study and are typically captured in the documents such as:

- P&IDs;
- SIF Philosophy document;
- Functional Logic Diagram.

Framework for the SRS

Prior to commencing any design work, the SRS must be prepared based on the guidance provided in IEC61511-1, 10. The SRS should contain both functional and integrity requirements for each SIF and should provide sufficient information to design and engineer the SIS. It should be expressed and structured to be clear, precise, verifiable, maintainable and feasible such as to aid comprehension by those who are likely to use the information at any phase in the lifecycle.

The SRS should include statements on the following for each SIF:

- Description of the SIF;
- Common cause failures;
- Safe state definition for the SIF;
- Demand rate;
- Proof test intervals;
- Response time to bring the process to a safe state;
- SIL and mode of operation (demand or continuous);
- Process measurements and their trip points;
- Process output actions and successful operation criteria;
- Functional relationship between inputs and outputs;
- Manual shutdown requirements;
- Energizing or de-energizing to trip;
- Resetting after a shutdown;

- Maximum allowed spurious trip rate;
- Failure modes and SIS response to failures;
- Starting up and restarting the SIS;
- Interfaces between the SIS and any other system;
- Application software;
- Overrides / inhibits / bypasses and how they will be cleared;
- Actions following a SIS fault detection.

Non-safety instrumented functions may be carried out by the SIS to ensure orderly shutdown or faster start-up.

11. RELIABILITY TECHNIQUES

Introduction

This section provides a brief introduction to reliability techniques. It is by no means a comprehensive survey of reliability engineering methods, nor is it in any way new or unconventional and the methods described herein are routinely used by reliability engineers.

Definitions

For convenience, an abridged list of key terms and definitions is provided. More comprehensive definitions of terms and nomenclature can be found in many standard texts on the subject.

Availability – A measure of the degree to which an item is in the operable and committable state at the start of the mission, when the mission is called for at an unknown state.

Failure – The event, or inoperable state, in which an item, or part of an item, does not, or would not, perform as previously specified.

Failure, dependent – Failure which is caused by the failure of an associated item(s). Not independent.

Failure, independent – Failure which occurs without being caused by the failure of any other item. Not dependent.

Failure mechanism – The physical, chemical, electrical, thermal or other process which results in failure.

Failure mode – The consequence of the mechanism through which the failure occurs, i.e. short, open, fracture, excessive wear.

Failure, random – Failure whose occurrence is predictable only in the probabilistic or statistical sense. This applies to all distributions.

Failure rate – The total number of failures within an item population, divided by the total number of life units expended by that population, during a particular measurement interval under stated conditions.

Maintenance, corrective – All actions performed, as a result of failure, to restore an item to a specified condition. Corrective maintenance can include any or all of the following steps: localization, isolation, disassembly, interchange, reassembly, alignment and checkout.

Maintenance, preventive – All actions performed in an attempt to retain an item in a specified condition by providing systematic inspection, detection and prevention of incipient failures.

Mean time between failures (MTBF) – A basic measure of reliability for repairable items: the mean number of life units during which all parts of the item perform within their specified limits, during a particular measurement interval under stated conditions.

Mean time to failure (MTTF) – A basic measure of reliability for non-repairable items: The mean number of life units during which all parts of the item perform within their specified limits, during a particular measurement interval under stated conditions.

Mean time to repair (MTTR) – A basic measure of maintainability: the sum of corrective maintenance times at any specified level of repair, divided by the total number of failures within an item repaired at that level, during a particular interval under stated conditions.

Reliability – (1) The duration or probability of failure-free performance under stated conditions. (2) The probability that an item can perform its intended function for a specified interval under stated conditions. For non-redundant items this is the equivalent to definition (1). For redundant items, this is the definition of mission reliability.

Basic mathematical concepts in reliability engineering

Many mathematical concepts apply to reliability engineering, particularly from the areas of probability and statistics. Likewise, many mathematical distributions can be used for various purposes, including the Gaussian (normal) distribution, the log-normal distribution, the Rayleigh distribution, the exponential distribution, the Weibull distribution and a host of others. For the purpose of this brief introduction, we'll limit our discussion to the exponential distribution.

Failure rate and mean time between/to failure (MTBF/MTTF).

The purpose of quantitative reliability measurements is to define the rate of failure relative to time and to model that failure rate in a mathematical distribution for the purpose of understanding the quantitative aspects of failure. The most basic building block is the failure rate, which is estimated using the following equation:

$$\lambda \quad = \quad F/T$$

Where: λ = Failure rate (sometimes referred to as the hazard rate);

T = total number of device hours (running time/cycles/miles/etc.) during an investigation period for both failed and non-failed items;

F = the total number of failures occurring during the investigation period.

For example, if five electric motors operate for a collective total time of 50 years with five functional failures during the period, the failure rate is 0.1 failures per year.

Another very basic concept is the mean time between/to failure (MTBF/MTTF). The only difference between MTBF and MTTF is that we employ MTBF when referring to items that are repaired when they fail. For items that are simply thrown away and replaced, we use the term MTTF. The computations are the same.

The basic calculation to estimate mean time between failure (MTBF) and mean time to failure (MTTF), is the reciprocal of the failure rate function. It is calculated using the following equation.

$$\theta \quad = \quad T/F$$

Where: θ = Mean time between/to failure;

T = Total running time/cycles/miles/etc. during an investigation period for both failed and non-failed items;

F = the total number of failures occurring during the investigation period.

The MTBF for our industrial electric motor example is 10 years, which is the reciprocal of the failure rate for the motors. Incidentally, we would estimate MTBF for electric motors that are rebuilt upon failure. For smaller motors that are considered disposable, we would state the MTTF.

The failure rate is a basic component of many more complex reliability calculations. Depending upon the mechanical/electrical design, operating context, environment and/or maintenance effectiveness, a machine's failure rate as a function of time may decline, remain constant, increase linearly or increase geometrically. However, for most reliability calculations, a constant failure rate is assumed.

The Bathtub Curve

In concept, the bathtub curve demonstrates a machine's three basic failure rate characteristics: decreasing, constant and increasing, Figure 39.

FIGURE 39. BATH TUB CURVE

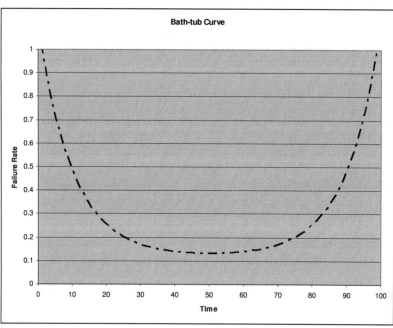

In practice, most machines spend their lives in the early life, or the constant failure rate regions of the bathtub curve. We rarely see time-dependent failures mechanisms as typical industrial machines tend to be replaced, or have parts replaced, before they wear out. However, despite its modeling limitations, the bathtub curve is a useful tool for explaining the basic concepts of reliability engineering.

There is a notion that the bathtub curve as a composite of several failure distributions, Figure 40.

FIGURE 40. BATH TUB CURVE

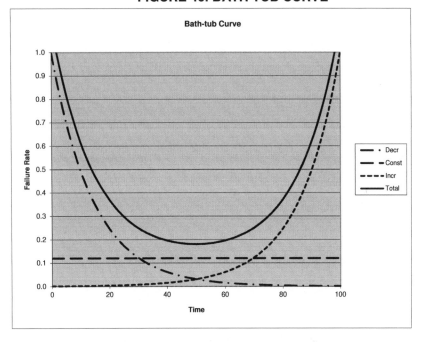

The human body is an excellent example of a system that follows the bathtub curve. People, and machines, tend to suffer a high failure rate (mortality) during their first years of life, but the rate decreases as the child (product) grows older. Assuming a person survives his or her teenage years, the mortality rate becomes fairly constant and remains there until age (time) dependent illnesses begin to increase the mortality rate (wearout).

The decreasing early life failure rate being due to systematic reasons such as manufactured weaknesses that are present in a product. As a batch of products is produced, a proportion of the population will contain weaknesses which will fail in service. As the failed items are returned for repair, the proportion of weak products in the population reduces and the failure rate decreases accordingly.

The increasing wear-out failures may be due to similar systematic reasons. Failure mechanisms may be due to degraded strength such as the accumulation of fatigue damage. In electronics, the time dependent failure mechanisms tend to be mechanical in nature and include fatigue failure of solder joints.

The constant failure rate period makes up the majority of a product's life and is a measure of the design quality, the goodness of the design. It is this constant failure rate region where simple reliability calculations are performed.

The Exponential Distribution

The exponential distribution, the most basic and widely used reliability prediction formula, models machines with the constant failure rate, or the flat section of the bathtub curve. Most industrial machines spend most of their lives in the constant failure rate, so it is widely applicable.

Below is the basic equation for estimating the reliability of a machine that follows the exponential distribution, where the failure rate is constant as a function of time.

$$R(t) \quad = \quad \exp \{ -\lambda \cdot t \}$$

Where: $R(t)$ = Reliability estimate for a period of time, cycles, miles, etc. (t); λ = Failure rate (1/MTBF, or 1/MTTF) and t = the time at risk.

In our electric motor example, if you assume a constant failure rate the likelihood of running a motor for six years without a failure, or the projected reliability, is 55 percent. This is calculated as follows:

$$
\begin{aligned}
R(t) \quad &= \quad \exp \{ - 0.1 \times 6 \} \\
&= \quad \exp \{ - 0.6 \} \\
&= \quad 0.5488 \\
&\approx \quad 55\%
\end{aligned}
$$

In other words, after six years, about 45% of the population of identical motors operating in an identical application can probabilistically be expected to fail. It is worth reiterating at this point that these calculations project the probability for a population. Any given individual from the population could fail on the first day of operation while another individual could last 30 years. That is the nature of probabilistic reliability projections.

A characteristic of the exponential distribution is the MTBF occurs at the point at which the calculated reliability is 36.78%, or the point at which 63.22% of the machines have already failed. In our motor example, after 10 years, 63.22% of the motors from a population of identical motors serving in identical applications can be expected to fail. In other words, the survival rate is 36.78% of the population.

Estimating System Reliability

Once the reliability of components or machines has been established relative to the operating context and required mission time, plant engineers must assess the reliability of a system or process. Again, for the sake of brevity and simplicity, we'll discuss system reliability estimates for series, parallel and shared-load redundant (M out of N) systems (MooN systems).

Series Systems

Before discussing series systems, we should discuss Reliability Block Diagrams (RBDs). RBDs simply map a process from start to finish. For a series system, Subsystem 1 is followed by Subsystem 2 and so forth. In the series system, the ability to employ Subsystem 2 depends upon the operating state of Subsystem 1. If Subsystem 1 is not operating, the system is down regardless of the condition of Subsystem 2 [Figure 41].

FIGURE 41. SERIES SYSTEM

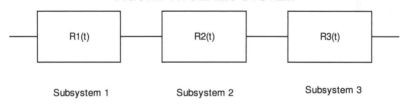

Subsystem 1   Subsystem 2   Subsystem 3

To calculate the system reliability for a serial process, you only need to multiply the estimated reliability of Subsystem 1 at time (t) by the estimated reliability of Subsystem 2 at time (t). The basic equation for calculating the system reliability of a simple series system is:

Rs(t)   =   R1(t) . R2(t) . R3(t)

Where: Rs(t) – System reliability for given time (t); Rn(t) – Subsystem or sub-function reliability for given time (t).

So, for a simple system with three subsystems, or sub-functions, each having an estimated reliability of 0.90 (90%) at time (t), the system reliability is calculated as 0.90 X 0.90 X 0.90 = 0.729, or about 73%.

Parallel Systems

Often, design engineers will incorporate redundancy into critical machines. Reliability engineers call these parallel systems. These systems may be designed as active parallel systems or standby parallel systems. The block diagram for a simple two component parallel system is shown in Figure 42.

FIGURE 42. PARALLEL SYSTEM

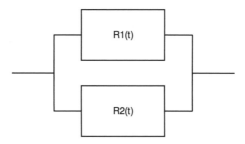

To calculate the reliability of an active parallel system, where both machines are running, use the following simple equation:

Rs(t)   =   1 – [{1-R1(t)} . {1-R2(t)}]

Where: Rs(t) – System reliability for given time (t); Rn(t) – Subsystem or sub-function reliability for given time (t).

The simple parallel system in our example with two components in parallel, each having a reliability of 0.90, has a total system reliability of 1 – (0.1 X 0.1) = 0.99. So, the system reliability was significantly improved.

M out of N Systems (MooN)

An important concept to plant reliability engineers is the concept of M out of N (MooN) systems. These systems require that M units from a total population of N be available for use. A good industrial example is coal pulverizers in an electric power generating plant. Often, the engineers design this function in the plant using an MooN approach.

For instance, if a unit has four pulverizers and requires that three of the four be operable to run at full load, then the system RBD would be for a 3oo4 configuration [Figure 43].

FIGURE 43. 3oo4 SYSTEM

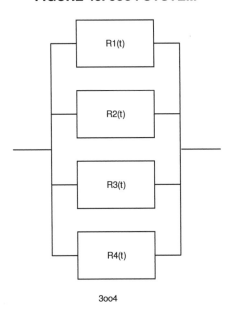

3oo4

Dangerous and Safe Failures

For reliability calculations to be meaningful, we are not only concerned with the failure rate of the system, but also how a system may fail, i.e. the failure mode.

Failure modes can be classified as safe or dangerous. Figure 44 shows a gas pipeline. If the pipeline is providing fuel to a power station and the Shutdown Valve fails and spuriously closes, then the fuel supply is cut off and perhaps there will be some loss of revenue but the failure mode (fail closed) is a **safe** failure.

If the same valve fails in the open position, then we maintain the fuel supply but should there be an overpressure condition, we will not be able to isolate the fuel and make the pipeline safe. This failure mode (fail open) is therefore considered a **dangerous** failure.

FIGURE 44. EXAMPLE SAFETY INSTRUMENTED FUNCTION

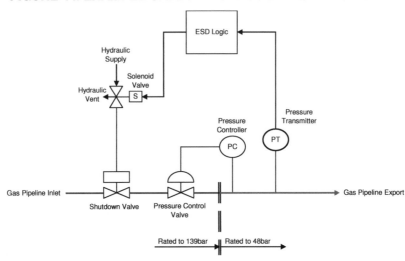

In this example, the fail open dangerous failure mode would not be revealed until a demand is placed upon it, i.e. until the valve is commanded to close. This is considered a dangerous undetected failure.

Alternatively, if the pipeline is providing a flow of coolant to the power station and the Shutdown Valve fails and spuriously closes, then the coolant is cut off and the power station may overheat. In this application, the same valve and the same failure mode (fail closed) is a **dangerous** failure. If the valve fails in the open position, then we maintain the coolant flow and therefore this failure mode (fail open) is therefore considered a **safe** failure.

A dangerous failure of a component in a safety instrumented function prevents that function from achieving a safe state when it is required to do so. **The dangerous failure rate is denoted by the symbol: λ_D.**

A safe failure does not have the potential to put the safety instrumented system in a dangerous or fail-to-function state but fails in such a way that it calls for the system to be shut down or the safety instrumented function to activate when there is no hazard present. **The safe failure rate is denoted by the symbol: λ_S.**

There may be failure modes which do not affect the safety function at all. These may include maintenance functions, indicators, data logging and other non-safety related (non-SR) functions. The non-SR failure rate is denoted by the symbol: $\lambda_{\text{non-SR}}$.

The total failure rate of an item, λ is equal to the sum of the safety-related and non-SR failure rates. Usually only λ_D and λ_S are included in reliability calculations.

$$\lambda = \lambda_D + \lambda_S + \lambda_{\text{non-SR}}$$

Detected and Undetected Failures

The PFD relates to dangerous failures that prevent the SIS from operating when required. These failure modes are either classified as detected failures, in that they are detected by diagnostics, or undetected failures that are not detected except by manual proof tests, which are typically performed annually. It is recommended that failure modes classified by Failure Modes, Effects and Criticality Analysis (FMECA), as dangerous detected failures should be detected as part of the diagnostics and verified in software validation. In addition, proof test procedures should ensure that dangerous undetected failure modes are revealed to ensure that proof tests are effective.

In accordance with IEC61508-6, Annex B.3.1, analysis may consider that for each safety function there is perfect proof testing and repair, i.e. all failures that are undetected are revealed by proof testing.

Proof Test Period (T_p) and Mean Down Time (MDT)

If a failure occurs, it is assumed that on average it will occur at the mid-point of the test interval. In other words, the fault will remain undetected for 50% of the test period.

For both detected and undetected failures the Mean Down Time (MDT) depends upon the test interval and also the repair time, or MTTR.

The MDT is therefore calculated from:

$$\text{MDT} = \frac{\text{test interval}}{2} + \text{MTTR}$$

The MDT for detected failures therefore approximates to the repair time, since the test interval (autotest) is generally short compared to the MTTR. For undetected failures the repair time is short compared to the test interval, the Proof Test period Tp, and therefore MDT for undetected failures approximates to Tp/2.

12. RELIABILITY BLOCK DIAGRAMS AND SYSTEM MODELLING

Modelling System Failure Rate

The failure rate of a redundant system λsys, can be calculated by considering the number of ways that system failure can occur. In a 3oo4 system, 3 out of the 4 channels are required to operate for the system to operate, therefore any two failures will result in a system failure, Figure 43.

FIGURE 45. 3oo4 SYSTEM

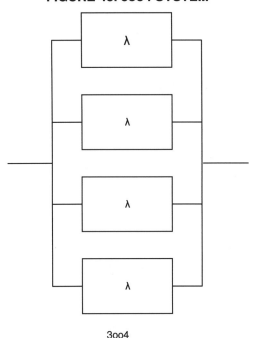

3oo4

The rate at which two failures can occur, λ_2 is given by the failure rate of one element λ, multiplied by the probability that a second failure will occur during the down time, MDT of the first failure, λ.MDT. Therefore:

$$\lambda_2 \quad = \quad \lambda.(\,\lambda.MDT\,)$$

However, there are 12 permutations (the order is important) of two failures in a 3oo4 system: A.B, A.C, A.D, B.C, B.D, C.D, B.A, C.A, D.A, C.B, D.B and D.C, and we must account for all of them. The system failure rate therefore becomes approximately:

$$\lambda_{SYS} \quad = \quad \textbf{12.}\lambda^2\textbf{.MDT}$$

To be exact, we should include all permutations of 3 and 4 concurrent failures as these will also result in system failure but as a first order approximation, these higher order terms can be neglected. The system failure rate for 3oo4, and other configurations are presented in Table 11.

Note, these are approximations that also neglect higher order terms.

TABLE 11. SYSTEM FAILURE RATE

Configuration	λsys
1oo1	λ
1oo2	$2.\lambda^2.MDT$
2oo2	$2.\lambda$
1oo3	$3.\lambda^3.MDT^2$
2oo3	$6.\lambda^2.MDT$
3oo3	$3.\lambda$
1oo4	$\lambda^4.MDT^3$
2oo4	$12.\lambda^3.MDT^2$
3oo4	$12.\lambda^2.MDT$
4oo4	$4.\lambda$

Modelling Dangerous Detected (λ_{DD}) and Dangerous Undetected Failure Rates (λ_{DU})

By substituting λ_{DD} and λ_{DU}, for λ in Table 11, and by using either the MDT or Tp/2 as appropriate, the system failure rate due to dangerous detected or undetected failures can be derived, Table 12.

TABLE 12. DANGEROUS SYSTEM FAILURE RATE

Configuration	Detected	Undetected
	λsys	λsys
1oo1	λ_{DD}	λ_{DU}
1oo2	$2.\lambda_{DD}^{2}.MDT$	$\lambda_{DU}^{2}.T_P$
2oo2	$2.\lambda_{DD}$	$2.\lambda_{DU}$
1oo3	$3.\lambda_{DD}^{3}.MDT^{2}$	$\lambda_{DU}^{3}.T_P^{2}$
2oo3	$6.\lambda_{DD}^{2}.MDT$	$3.\lambda_{DU}^{2}.T_P$
3oo3	$3.\lambda_{DD}$	$3.\lambda_{DU}$
1oo4	$\lambda_{DD}^{4}.MDT^{3}$	$\lambda_{DU}^{4}.T_P^{3}$
2oo4	$12.\lambda_{DD}^{3}.MDT^{2}$	$4.\lambda_{DU}^{3}.T_P^{2}$
3oo4	$12.\lambda_{DD}^{2}.MDT$	$6.\lambda_{DU}^{2}.T_P$
4oo4	$4.\lambda_{DD}$	$4.\lambda_{DU}$

Modelling System Spurious Trip Rate (λ_{STR})

In a series (1oo1) configuration, safe failures generally result in a system trip and they can therefore be assumed to be detected. In a redundant configuration, the system spurious trip rate will depend upon the redundancy and the voting implemented.

If a redundant channel fails and, because of the voting, the system does not trip, then the channel failure will be detected and repaired. The approach taken in modelling dangerous detected failures will therefore apply except that the number of failures required for a spurious trip may differ from that required for a dangerous failure.

Usually spurious trips include the safe failures only but depending on system failure behaviour on detection of a fault, dangerous detected failures may also be included and the spurious trip rate will be the sum of the two.

Table 13 summarises the system spurious trip rates for safe failures.

TABLE 13. SPURIOUS SYSTEM TRIP RATE

Configuration	Spurious
	λstr
1oo1	λ_S
1oo2	$2.\lambda_S$
2oo2	$2.\lambda_S^2.MDT$
1oo3	$3.\lambda_S$
2oo3	$6.\lambda_S^2.MDT$
3oo3	$3.\lambda_S^3.MDT^2$
1oo4	$4.\lambda_S$
2oo4	$12.\lambda_S^2.MDT$
3oo4	$12.\lambda_S^3.MDT^2$
4oo4	$\lambda_S^4.MDT^3$

Modelling Demand Mode Safety System Availability

For a safety system, the availability due to dangerous detected failures, A_{DD} is given by:

$$A_{DD} \quad = \quad 1 / (1 + .\lambda_{DD(SYS)}.MDT)$$

where $\lambda_{DD(SYS)}$ is the system failure rate as a result of dangerous detected failures.

For dangerous undetected failures, A_{DU} is given by:

$$A_{DU} \quad = \quad 1 / (1 + .\lambda_{DU(SYS)}.T_P / 2)$$

where $\lambda_{DU(SYS)}$ is the system failure rate as a result of dangerous undetected failures.

For safe failures, A_S is given by:

$$A_S \quad = \quad 1 / (1 + .\lambda_{S(SYS)}.MDT)$$

where $\lambda_{S(SYS)}$ is the system failure rate as a result of spurious (safe) failures.

System availability is the product of the availabilities due to dangerous detected, dangerous undetected and safe failures:

$$A_{SYS} \quad = \quad A_{DD} \cdot A_{DU} \cdot A_{S}$$

This method can be used for modelling series (simplex) systems and also redundant systems.

Modelling Continuous Mode Safety System Availability

When the method is applied to continuous mode safety systems, the analyst must understand the nature of demands placed upon the safety function. Some continuous mode safety functions operate on demand (just as a demand mode safety function) but they are classed as continuous mode because of the demand frequency, i.e. greater than once per year. In this case, availability can be calculated as for a demand mode safety function except that the proof test interval T_P should be replaced with the demand interval, T_D. Dangerous undetected failures would be unrevealed until a demand is placed upon the safety function.

Where the continuous mode safety function effectively provides continuous control then the availability can be calculated as a control system.

Modelling Control System Availability

In modelling the availability of Control Systems we are concerned about failures that affect the process and we must decide whether a failure affects the process to such an extent that the control system is effectively unavailable.

Detection of a failure will be either by diagnostics and fault alarms, in which case a repair is required and the system will be unavailable until it is restored, or by symptom, in which case the process under control operates out of set point limits.

Failures that are undetected do not immediately result in the control system being unavailable. In time, the undetected failure may result in a process parameter drifting out of specified limits at which time, it will be revealed and result in unavailability.

Control system availability can therefore be modelled by considering the total system failure rate, A_{SYS} is given by:

$$A_{SYS} = 1 / (1 + \lambda_{SYS}.MDT)$$

where λ_{SYS} is the total system failure rate as a result of all failures [Table 11].

Probability of Dangerous Failure/Hour (PFH) and Probability of Failure on Demand (PFD)

The simplified PFH and PFD formulae for common configurations are presented in Table 14, for detected failures and in Table 15, for undetected failures.

TABLE 14. PFH / PFD CALCULATION (DETECTED FAILURES)

Configuration	PFH	PFD
1oo1	λ_{DD}	$\lambda_{DD}.MDT$
1oo2	$2.\lambda_{DD}^{2}.MDT$	$\lambda_{DD}^{2}.MDT^{2}$
2oo2	$2.\lambda_{DD}$	$2.\lambda_{DD}.MDT$
1oo3	$3.\lambda_{DD}^{3}.MDT^{2}$	$\lambda_{DD}^{3}.MDT^{3}$
2oo3	$6.\lambda_{DD}^{2}.MDT$	$3.\lambda_{DD}^{2}.MDT^{2}$
3oo3	$3.\lambda_{DD}$	$3.\lambda_{DD}.MDT$
1oo4	$4.\lambda_{DD}^{4}.MDT^{3}$	$\lambda_{DD}^{4}.MDT^{4}$
2oo4	$12.\lambda_{DD}^{3}.MDT^{2}$	$4.\lambda_{DD}^{3}.MDT^{3}$
3oo4	$12.\lambda_{DD}^{2}.MDT$	$6.\lambda_{DD}^{2}.MDT^{2}$
4oo4	$4.\lambda_{DD}$	$4.\lambda_{DD}.MDT$

TABLE 15. PFH / PFD CALCULATION (UNDETECTED FAILURES)

Configuration	PFH	PFD
1oo1	λ_{DU}	$\lambda_{DD}.T_P/2$
1oo2	$\lambda_{DU}{}^2.T_P$	$\lambda_{DD}{}^2.T_P{}^2/3$
2oo2	$2.\lambda_{DU}$	$\lambda_{DD}.T_P$
1oo3	$\lambda_{DU}{}^3.T_P{}^2$	$\lambda_{DD}{}^3.T_P{}^3/4$
2oo3	$3.\lambda_{DU}{}^2.T_P$	$\lambda_{DD}{}^2.T_P{}^2$
3oo3	$3.\lambda_{DU}$	$3.\lambda_{DD}.T_P/2$
1oo4	$\lambda_{DU}{}^4.T_P{}^3$	$\lambda_{DD}{}^4.T_P{}^4/5$
2oo4	$4.\lambda_{DU}{}^3.T_P{}^2$	$\lambda_{DD}{}^3.T_P{}^3$
3oo4	$6.\lambda_{DU}{}^2.T_P$	$2.\lambda_{DD}{}^2.T_P{}^2$
4oo4	$4.\lambda_{DU}$	$2.\lambda_{DD}.T_P$

Accounting for Common Cause Failures

Common cause failures (CCF) are failures that may result from a single cause but simultaneously affect more than one channel. They may result from a systematic fault for example, a design specification error or an external stress such as an excessive temperature that could lead to component failure in both redundant channels. It is the responsibility of the system designer to take steps to minimise the likelihood of common cause failures by using appropriate design practices.

The contribution of CCF in parallel redundant paths is accounted for by inclusion of a β-factor. The CCF failure rate that is included in the calculation is equal to β x the total failure rate of one of the redundant paths.

The β-factor model [IEC61508-6, Annex D] is the preferred technique because it is objective and provides traceability in the estimation of β. The model has been compiled to ask a series of specific questions, which are then scored using objective engineering judgement. The maximum score for each question has been weighted in the model by calibrating the results of various assessments, against known field failure data.

Two columns are used for checklist scores. Column A contains the scores for those features of CCF protection that are perceived as being enhanced by an increase of diagnostic frequency (auto-test or proof test). Column B contains the scores for those features thought not to be enhanced by an improvement in diagnostic frequency.

The model allows the scoring to be modified by the frequency and coverage of diagnostic test. Column A scores are multiplied by a factor C, which is derived from diagnostic related considerations. The final β-factor is then estimated from the raw score total:

$$\text{Raw score} \quad = \quad (A * C) + B$$

The relationship between β and the raw score is essentially a negative exponential function, since there is no data to justify departure from the assumption that as β decreases (improves) then successive improvements become increasingly more difficult to achieve.

Where a particular question may not apply to the system being evaluated, a score of either 100% or 0% is entered depending upon which is appropriate for the system.

The following represents typical constraints that may be considered for the purposes of estimating CCF contribution:

- redundant channels are physically separated;
- diverse technologies, e.g. one electronic channel and one relay based channel;
- written system of work on site should ensure that failures are investigated;
- written maintenance procedures should prevent re-routing of cable runs;
- personnel access is limited;
- the operating environment is controlled and the equipment has been rated over the full environmental range.

Actual in-service performance however, will depend upon the specific installation and the design, operating and maintenance practices that are adopted but provided that all appropriate good engineering practices are adopted, then the model will provide a traceable estimation of CCF contribution.

When taking CCFs into account in the formulae for PFD and PFH [Table 14 and Table 15] the following approach can be used. The equations used are simplifications to the standard equations and are derived in [20.8].

The following provides a summary of the basic forms.

For detected failures:

$$PFD_{1oo1} = \lambda_{DD} . MDT$$

Ref. IEC61508-6, B.3.2.2.1

$$PFD_{1oo2} = \lambda_{DD}^2 . MDT^2 + \beta . \lambda_{DD} . MDT$$

Ref. IEC61508-6, B.3.2.2.2

For undetected failures:

$$PFD_{1oo1} = \lambda_{DU} . T_p / 2$$

Ref. IEC61508-6, B.3.2.2.1

$$PFD_{1oo2} = \lambda_{DU}^2 . T_p^2 / 3 + \beta . \lambda_{DU} . T_p / 2$$

Ref. IEC61508-6, B.3.2.2.2

Where λ_{DD} is the dangerous detected failure rate, λ_{DU} is the dangerous undetected failure rate and β is the contribution from common cause failures. T_p is the proof test interval and MDT is the Mean Down Time.

The generic forms of these equations for various configurations, for both continuous mode and demand mode systems are examined in [20.9].

Failure Rates

In the calculation of PFD and SFF, the analysis uses the underlying hypothesis of IEC61508-6, Annex B.3 in that component failure rates are constant over the lifetime of the system.

The failure rates used in calculations may be obtained by Failure Modes, Effects and Criticality Analysis (FMECA), quantified by field data, or by reference to published data from industry sources. The failure rates used should be compared with available data for similar modules of complexity and technology. This approach ensures a conservative approach in terms of reliability modelling and gives confidence that the calculated reliability performance should be achievable in service.

Example Data Sheet

The failure rate data used in the previous RBDs should be visible in the report and should show traceability to source. The source, where referring to published data, should show enough detail that would allow third parties to independently verify the data used. This may include document identification, ISBN number if applicable and page and item number.

Table 16 shows a typical data table for the previous example RBDs.

TABLE 16. DANGEROUS SYSTEM FAILURE RATE

Description	λTotal	λD	λDD	λDU	λS	Comments / Source
Pressure Transmitter PT-xxx	3.7E-06	1.5E-06	1.2E-06	3.7E-07	2.2E-06	Manufacturers PT-xxx Functional Safety Manual.
Fan Loading FL-xxx Current Transformer	5.0E-07	2.0E-07	0.0E+00	2.0E-07	3.0E-07	FARADIP-THREE V6.4, Reliability Data Base. Technis.
Comms. Module ControlNet CNB	1.8E-07	9.1E-08	8.2E-08	9.1E-09	9.1E-08	Allen-Bradley ControlLogix in SIL2 Applications.
Analogue Input Module	6.6E-07	3.3E-07	2.9E-07	3.3E-08	3.3E-07	Allen-Bradley ControlLogix in SIL2 Applications.
ControlLogix CPU	4.5E-07	2.3E-07	2.0E-07	2.3E-08	2.3E-07	Allen-Bradley ControlLogix in SIL2 Applications.
Digital Output Module	3.1E-07	1.5E-07	1.4E-07	1.5E-08	1.5E-07	Allen-Bradley ControlLogix in SIL2 Applications.

Modelling 1oo2, 1oo2D and Hot Standby

The following examples show RBDs modelling some common system configurations.

Modelling 1oo2

A 1oo2 system is a 1 out of 2 architecture, where either of the two channels can perform the safety function. This is a fault tolerant configuration where one channel failure can be tolerated.

If the channel failure is a dangerous unrevealed failure, this will not be detected by the diagnostics and there will be no fault indication. However the safety function will still operate as the 1 remaining channel can initiate the trip. If the channel failure is a dangerous detected failure, this will usually result in a fault indication.

Modelling 1oo2D

A 1oo2D system architecture has two channels connected in parallel and each channel has diagnostic circuits to detect failures with a high diagnostic coverage. Both channels need to agree to execute a shutdown action during normal operation of the system. A healthy channel controls the system if the diagnostic circuit of the other side detects a failure.

In terms of reliability modelling, for dangerous detected failures, the 1oo2D system operates as a 1oo2 configuration and system failure rate and PFD can be modelled as a 1oo2 for detected failures. However, a single dangerous undetected channel failure in a 1oo2D system will prevent the system from operating and therefore system failure rate and PFD must be modelled as 2oo2 for undetected failures. In other words, both channels must operate.

Modelling Hot standby

A hot standby system architecture has two channels connected in parallel such that one channel is designated a master and controls the safety function. The other channel acts as a hot spare such that if a dangerous failure is detected in the master channel, then the standby channel takes over control of the safety function.

In terms of reliability modelling, for dangerous detected failures, the hot standby system operates as a 1oo2 configuration and system failure rate and PFD can be modelled as a 1oo2 for detected failures.

A single dangerous undetected channel failure will prevent the system from operating and therefore system failure rate and PFD must be modelled as 1oo1 for undetected failures. In other words, the safety function cannot tolerate an undetected failure of the master channel and there is no redundancy for undetected failures.

System failure rate for a 1oo2 System

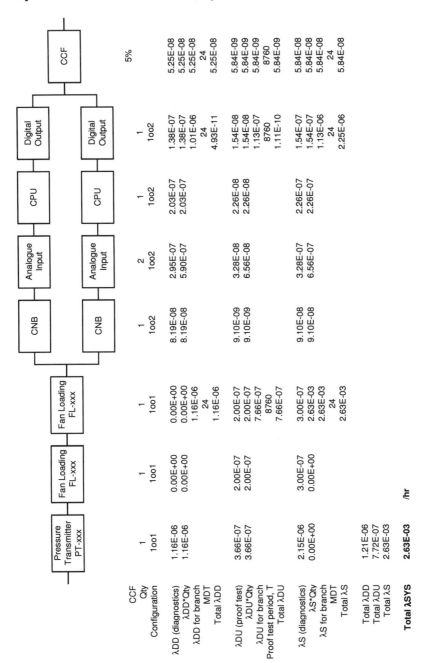

	Pressure Transmitter PT-xxx	Fan Loading FL-xxx	Fan Loading FL-xxx	CNB	Analogue Input	CPU	Digital Output	CCF
CCF								
Qty	1	1	1	1	2	1	1	
Configuration	1oo1	1oo1	1oo1	1oo2	1oo2	1oo2	1oo2	5%
λ_{DD} (diagnostics)	1.16E-06	0.00E+00	0.00E+00	8.19E-08	2.95E-07	2.03E-07	1.38E-07	5.25E-08
λ_{DD}*Qty	1.16E-06	0.00E+00	0.00E+00	8.19E-08	5.90E-07	2.03E-07	1.38E-07	5.25E-08
λ_{DD} for branch			1.16E-06				1.01E-06	5.25E-08
MDT			24				24	24
Total λ_{DD}			1.16E-06				4.93E-11	5.25E-08
λ_{DU} (proof test)	3.66E-07	2.00E-07	2.00E-07	9.10E-09	3.28E-08	2.26E-08	1.54E-08	5.84E-09
λ_{DU}*Qty	3.66E-07	2.00E-07	2.00E-07	9.10E-09	6.56E-08	2.26E-08	1.54E-08	5.84E-09
λ_{DU} for branch			7.66E-07				1.13E-07	5.84E-09
Proof test period, T			8760				8760	8760
Total λ_{DU}			7.66E-07				1.11E-10	5.84E-09
λ_{S} (diagnostics)	2.15E-06	3.00E-07	3.00E-07	9.10E-08	3.28E-07	2.26E-07	1.54E-07	5.84E-08
λ_{S}*Qty	0.00E+00	0.00E+00	2.63E-03	9.10E-08	6.56E-07	2.26E-07	1.54E-07	5.84E-08
λ_{S} for branch			2.63E-03				1.13E-06	5.84E-08
MDT			24				24	24
Total λ_{S}			2.63E-03				2.25E-06	5.84E-08
Total λ_{DD}	1.21E-06							
Total λ_{DU}	7.72E-07							
Total λ_{S}	2.63E-03							
Total λ_{SYS}	**2.63E-03**	/hr						

System failure rate for a 1oo2D System

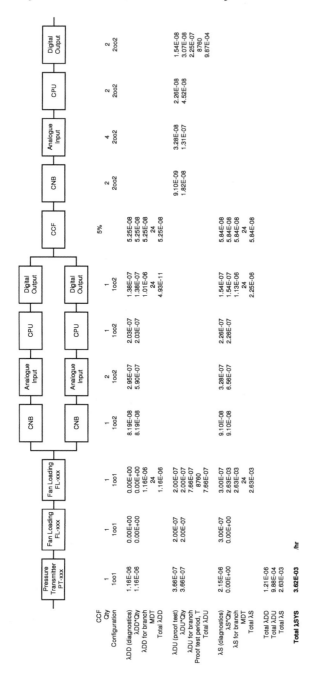

System failure rate for a Hot Standby System

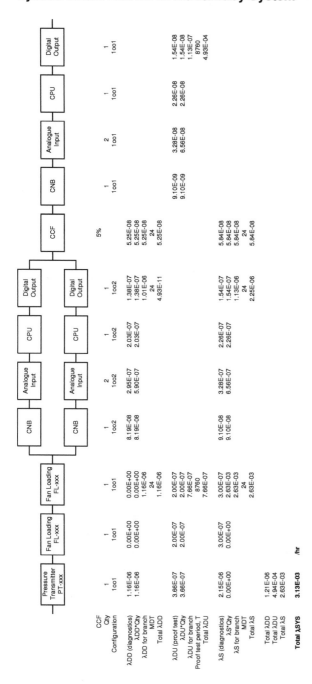

Availability of a Complex System

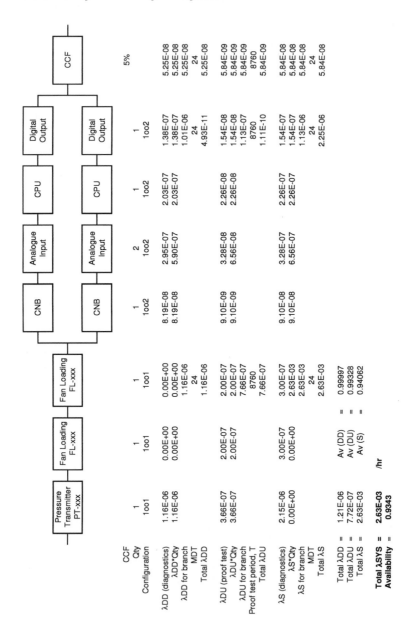

Modelling Fire and Gas (F&G) Systems

In modelling F&G Systems, it is important to give some guidance on Fault Tolerance. Modelling of ESD or similar systems typically follows the same configuration used by the logic solver voting. For example, the reliability of Pressure Transmitters which are voted one out of two, (1oo2) on high pressure by an ESD System, will be modelled as 1oo2. The same is not always true for F&G Systems.

In general, a conservative analysis can usually be undertaken without relying on assumptions of detector coverage and redundancy in alarm layout but in practice this may result in a pessimistic analysis and failure to meet targets. Where such difficulties arise, a detailed knowledge of the hazards allows a more targeted model to be developed enabling a more realistic reliability analysis to be carried out.

F&G Systems not only protect people, but they can also be used to protect an asset against commercial risk; or a site against an environmental risk and the executive action required by the SIF in providing this protection, will determine the appropriate reliability model to use.

When modelling a F&G SIF to determine compliance against hardware reliability targets, e.g. PFD, decisions must be taken to determine exactly what configuration of hardware to model.

As an example, C&E data for a F&G SIF will typically specify:

a) any one out of six (1oo6) gas detectors in the alarm state is denoted Single Gas and will activate a Control Room alarm;

b) any two out of six (2oo6) gas detectors in the alarm state is denoted Confirmed Gas and will activate site alarms and beacons, and generate an ESD of the plant.

However, for correct modelling, we must understand the SIF and the hazard it protects against. The executive action required by the SIF will determine the appropriate model to use.

Modelling F&G System Detector Configurations

In practice, a single gas alarm will be investigated by an operator to determine whether it is real, spurious, or due to a detector fault. Executive action is taken as a result of Confirmed Gas only and this will ensure that plant personnel evacuate to safety. This is the safety function that has attracted the SIL target and therefore case b) above should be our starting point for reliability modelling: a Confirmed Gas Alarm will ensure personnel evacuate to safety.

The layout in Figure 46 shows six Gas Detectors positioned in a zone, and the logic solver voting 2oo6, is configured to take executive action if any 2 out of the 6 detectors sense gas.

FIGURE 46. F&G SYSTEM LAYOUT

Zone with 6 Gas Detectors

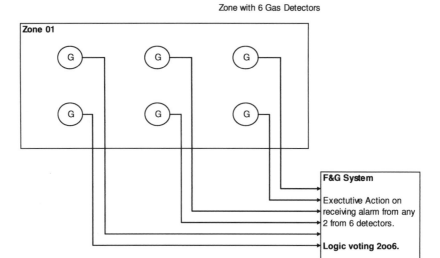

However, modelling SIFs against PFD targets is about calculating the probability of not reacting to gas when required. A gas release, which is large enough to be hazardous, may only be within the coverage of say, half of the 6 detectors, Figure 47.

FIGURE 47. F&G SYSTEM COVERAGE

Zone with 6 Gas Detectors

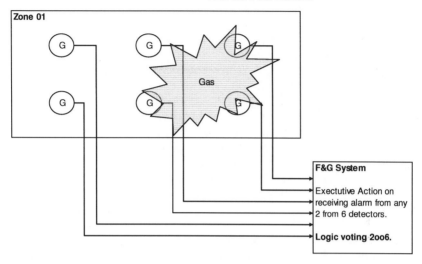

In practice, we are likely to require executive action to be initiated as soon as possible, i.e. when the minimum two sensors are within the gas cloud. In this case we should model the sensors as 2oo2, without redundancy, and consequently no sensor failures could be tolerated. If the targets are achieved with a non-redundant configuration, then this would represent a conservative approach because it does not rely on the justification of any assumptions of detector coverage.

In reality, the PFD of the sensor subsystem is likely to be better than that calculated for a non-redundant configuration because there is likely to be some overlap in sensor coverage due to their placement, and a single sensor failure could possibly be tolerated.

In terms of reliability modelling, the analyst must therefore judge the maximum size of gas release (cloud size) that could be tolerated before executive action is required and estimate how many sensors will fall within the cloud at this time.

In this example, if we can allow the gas cloud to be large enough to cover 3 sensors before we initiate executive action, then with the logic voting any 2 out of 6, we can tolerate one sensor failure. In other words, the reliability of the gas detection could be modelled as 2 out of 3.

Effect of Incorrect Modelling on PFD

In the above example, as the logic voting of gas detectors is 2oo6, some analysts succumb to the temptation to model the reliability of the system as 2oo6 instead of 2oo3 or even 2oo2. Obviously, the resulting discrepancy in the overall PFD of the safety function and its performance against SIL targets between redundant and non-redundant configurations can be significant.

Provided some fault tolerance can be reasonably claimed, e.g. by modelling 2oo3, or 2oo4 then the resulting differences in the overall PFD of the safety function and its performance against SIL targets will be small. The PFD for redundant configurations is limited by common cause failures and so improvements in PFD are not significant when the Hardware Fault Tolerance (HFT) increases above 1.

However, if the HFT cannot be assured due to either the detector placement or the size of gas cloud that can be tolerated when executive action is required, then the resulting discrepancy between redundant and non-redundant configurations, can be significant, Figure 48.

FIGURE 48. F&G SYSTEM PFD CALCULATION

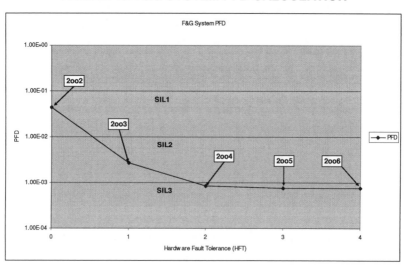

Note: PFD is calculated for typical sensors failure rates and repair times and assumes a contribution from common causes for redundant configurations. A fault tolerance of zero in this example represents a 2oo2 configuration, a fault tolerance of 1 represents 2oo3, 2 represents 2oo4 and so on.

The results show that depending upon the architecture, or HFT selected for modelling, the calculated PFD could fall within the SIL1, SIL2 or SIL3 band.

Effect of Incorrect Modelling on Architecture

Incorrect modelling will have a more significant effect on the architectural performance of the safety function. For a given Safe Failure Fraction (SFF) the SIL performance of the detector subsystem is dependent upon its HFT.

For example, for a Type B detector with a SFF of between 60% and 90%, the following architectural SIL capabilities could be claimed:

HFT	Configuration	SIL (Architecture)
0	2oo2	SIL1
1	2oo3	SIL2
2	2oo4	SIL3

Again, if the analyst assumes a 2oo6 configuration because of the voting logic, then an optimistic architecture would result in a SIL3 claimed when actually a lower SIL may only apply.

Modelling F&G System Alarm Configurations

Personnel are protected from fire and gas hazards by a Confirmed Alarm. The visible and audible alarms are all that is required to ensure personnel evacuate to safety. Therefore, for safety hazards, the output configuration needs only to consider the provision of visible and audible annunciators.

For F&G Systems, executive action may be typically specified as activating 6oo6 visual AND 4oo4 audible alarms. Modelling such configurations usually causes a problem in achieving anything better than a SIL1 PFD target because of the number of devices to be included. In addition, because alarms and beacons have a very low SFF, their architectural performance cannot usually achieve better than SIL1 in simplex configurations.

Bearing in mind, a zone may contain noisy equipment that may obscure a beacon or prevent an audible alarm from being heard, good practice should aim to position alarms so that personnel in the hazardous area can always see or hear more than one annunciator at a time. If this assumption can be verified then the analyst can take advantage of such fault tolerance in the reliability modelling of the alarm configuration.

A 6oo6 configuration of annunciators may cover 2 or 3 separate zones with maybe 2 or 3 annunciators per zone. The analyst must therefore decide from plant layout drawings, what fault tolerance can be claimed for each zone and then model this accordingly, Figure 49.

FIGURE 49. EXAMPLE ALARM SYSTEM LAYOUT

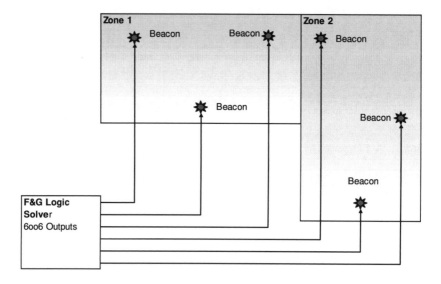

The key is to decide how many beacons can be seen and how many of those can be allowed to fail without causing the loss of the safety function. In the case study layout, it was decided that in each zone, two beacons out of the 3 in the zone, could always be seen.

In such a layout, a reasonable approach would be to model each zone 1 as 1oo2, because you only need to see one beacon. However, because both zones have to be protected, both zones would have to be included in the model, i.e. 1oo2 + 1oo2.

As a further example, consider 6 beacons in a single zone where, it was decided that at any time, 4 of the 6 beacons could be seen, Figure 50. Then one beacon is required to work out of the 4 that can be seen, and so we can model the alarms as 1oo4.

FIGURE 50. EXAMPLE ALARM SYSTEM LAYOUT (1 ZONE)

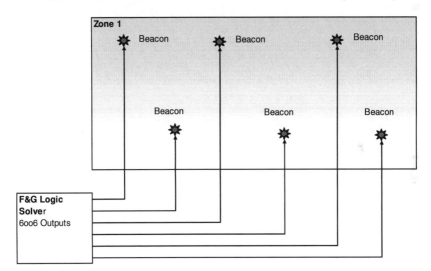

F&G Inputs to ESD Systems

So far, no mention has been made of the requirement, on Confirmed Fire or Gas, to generate an ESD of the plant. Whether to include the ESD Trip as part of the F&G SIF will depend upon the consequences of the hazard and the protection required.

Where the hazard results in a personal safety risk, it may be argued that alarms are sufficient to ensure protection. Usually, F&G trips also generate an input to the ESD but in many cases this is to prevent escalation of the hazard and to protect the asset. An ESD trip may also be initiated as good house-keeping, enabling start-up to be accomplished in a more controlled fashion following resolution of the hazard. The F&G System is there to protect against fire or gas; the ESD is there to protect against other hazards. Provided the F&G System meets its targets in terms of risk reduction, there should be no reason, other than outlined above, to trip the ESD. Therefore the ESD would not normally be included in the F&G SIF.

There are exceptions however. When the hazard leads to environmental or asset damage, alarms alone will provide no protection and therefore it may be necessary to isolate the plant on detection of fire or gas. In such cases, it is necessary to include shutdown and isolation as required in the reliability modelling of the F&G SIFs.

Summary

It can be seen that input subsystem modelling can give optimistic results if the logic voting configuration is modelled rather than the detector fault tolerance. The same approach when modelling the output subsystem will give very pessimistic results. Between the two subsystems, the modelling approach adopted can result in a large variation in calculated PFD and architectural performance and hence, a large variation in the SIL claimed is possible.

It is therefore important that a thoughtful approach to modelling F&G Systems is adopted and a clear understanding of modelling techniques, and the hazards and systems analysed, is acquired. This will ensure that an accurate assessment of the risk reduction provided by a F&G System is achieved and end users are not misinformed by optimistic claims.

13. SIL VERIFICATION

Complying with Safety Integrity Level Targets

Many people ask what they need to do to demonstrate compliance. It is not enough to buy SIL certified components and assume this will achieve out-of-the-box compliance, and, since the standard is non-prescriptive, it is also not possible to provide a checklist or tick-chart of what must be done. In truth, how much or how little you do will depend upon many things. The approach will depend upon how much information or data is available, the depth of analysis, or rigour applied must satisfy your customer or regulator but above all you must satisfy yourself that you have done enough.

If something goes wrong and someone is killed, can you face the families and demonstrate that you did everything that would be reasonably expected of you?

A suggested Compliance Plan would be to meet the requirements of IEC61511-1, 10 and 12. These include the following sub-clauses, as shown in Figure 51:

- Requirements for system behaviour on detection of a fault (IEC61511-1, 11.3);

- Hardware fault tolerance (IEC61511-1, 11.4);

- Selection of components and subsystems (IEC61511-1, 11.5);

- Field devices (IEC61511-1, 11.6);

- Operator, maintainer and communication interfaces with the SIS (IEC61511-1, 11.7);

- Maintenance or testing design requirements (IEC61511-1, 11.8);

- SIF probability of failure (IEC61511-1, 11.9);

- Application software (IEC61511-1, 11.12).

Where these clauses are further sub-divided into more detailed requirements, these are also shown.

Compliance with the IEC61511-1, 5: Management of Functional Safety is further discussed in section [19].

FIGURE 51. PLAN FOR COMPLIANCE

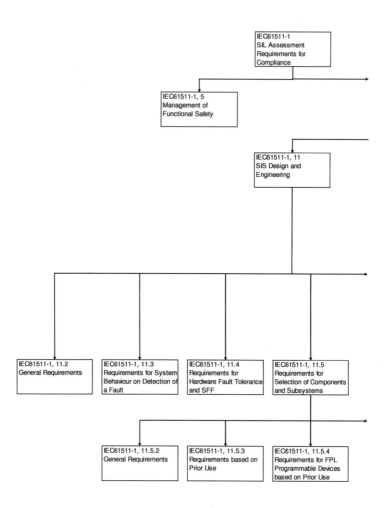

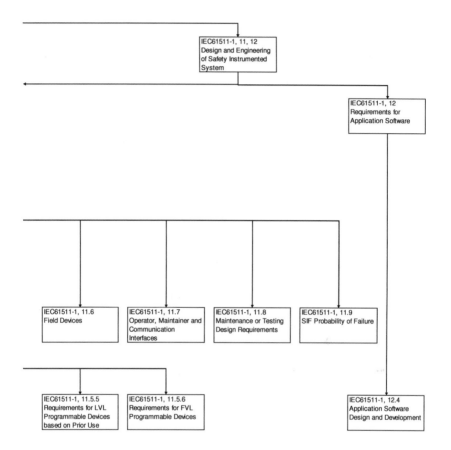

Requirements for system behaviour on detection of a fault IEC61511-1, 11.3

The system behaviour on detection of a fault should be specified. This may be detailed in the SRS or design specification for example.

The following shows *typical examples* of the kind of parameters that could be considered for inclusion:

1.   All output blocks vote 1oo2 on PLC demands and revert to 1oo1 on detection of loss of communications from a PLC.

2.   The design specification states that a fail-safe principle applies. All shutdown elements of the SIS achieve a failure to safety principle.

3.   In the case of an ESD System, a de-energise-to-trip function has been implemented.

4.   In the case of the F&G System, an energise-to-trip release of extinguishant has been implemented. The detection of a single dangerous fault in a redundant configuration is indicated by an alarm condition. The F&G System continues to operate safely for the allowed duration of the repair time and other additional risk reduction measures have been implemented such as the provision of a hardwired manual release of extinguishant.

Requirements for Hardware Fault Tolerance, IEC61511-1, 11.4

Approach

To address the requirements for Hardware Fault Tolerance (HFT), a quantitative assessment against Safe Failure Fraction (SFF) and Architectural Constraints is required.

Safe Failure Fraction

In the context of hardware safety integrity, the highest SIL that can be claimed for a safety function is limited by the HFT and the SFF, of the sub-systems that carry out that safety function.

A hardware fault tolerance of 1 indicates that the architecture of the sub-system is such that a dangerous failure of one of the sub-systems does not prevent the safety action from occurring i.e. a configuration of 1oo2 or 2oo3 would have a HFT of 1 or a configuration of 1oo3 or 2oo4 would have a HFT of 2.

With respect to these requirements, IEC61508 [20.1], gives the following additional guidance:

- a hardware fault tolerance of N means that N+1 faults could cause the loss of the safety function. In determining the hardware fault tolerance, no account shall be taken of other measures that may control the effects of faults such as diagnostics;

- where one fault directly leads to the occurrence of one or more subsequent faults, these are considered as a single fault;

- in determining hardware fault tolerance, certain faults may be excluded, provided that the likelihood of them occurring is very low in relation to the safety integrity requirements of the subsystem. Any such fault exclusions shall be justified and documented.

The following general relationships are used.

$$SFF = \Sigma \, (\Sigma \, \lambda_S + \Sigma \, \lambda_{DD}) \, / \, (\Sigma \, \lambda_S + \Sigma \, \lambda_D) \qquad \text{Ref. IEC61508-2.C.1}$$

where:

$$\lambda_D = \lambda_{DU} + \lambda_{DD}$$

For each element in the safety function, SFF should be calculated. The value should then be used in Table 17 to determine SIL compliance for the level of hardware fault tolerance.

Architectural Constraints

IEC61511-1, 11.4.5 allows for assessment of hardware fault tolerance using the requirements of IEC61508-2, Tables 2 and 3.

Within IEC61508 [20.1], subsystems are categorised as either Type A or Type B. Generally, if the failure modes are well defined and the behaviour under fault conditions can be completely determined and there is sufficient adequate field data then the subsystem may be considered Type A. If any of these conditions may not be true then the subsystem must be considered Type B.

Simple mechanical devices such as valves are generally considered to be Type A. Logic Solvers are usually Type B in that they contain some processing capability and as such, their behaviour under fault conditions may not be completely determined. Sensors can be either Type A or Type B depending on the technology and complexity of the device.

The architectural constraints for a safety function are summarised in Table 17.

TABLE 17. ARCHITECTURAL CONSTRAINTS

Type A subsystems definition:		
Failure modes of all constituent parts well defined, and		
Behaviour of the subsystem under fault conditions completely determined, and		
Sufficient dependable data from field experience to show that the claimed failure rates for detected and undetected dangerous failures are met		

Safe Failure Fraction	Hardware Fault Tolerance (N)		
	0	1	2
<60%	SIL 1	SIL 2	SIL 3
60% - <90%	SIL 2	SIL 3	SIL 4
90% - <99%	SIL 3	SIL 4	SIL 4
≥ 99%	SIL 3	SIL 4	SIL 4

Type B subsystems definition:		
Failure mode of at least one constituent component is not well defined, or		
The behaviour of the subsystem under fault conditions cannot be completely determined, or		
There is insufficient dependable data from field experience to support the claimed failure rates for detected and undetected dangerous failures		

Safe Failure Fraction	Hardware Fault Tolerance (N)		
	0	1	2
<60%	Not allowed	SIL 1	SIL 2
60% - <90%	SIL 1	SIL 2	SIL 3
90% - <99%	SIL 2	SIL 3	SIL 4
≥ 99%	SIL 3	SIL 4	SIL 4

Note: A hardware fault tolerance of N means that N+1 faults could cause the loss of the safety function.

Example

In this example, Figure 52 the safety function consists of two level switches operating in a 1oo2 configuration. If either switch detects a high level, then the PLC will de-energise the SOV which will allow the ESD Valve to close.

The assessment of the architectural performance requires that we first identify what type, i.e. A or B, each element is. This can generally be determined using the definitions provided in Table 17. As a general rule, you have to be sure of the failure modes and failure behaviour of an item, and have very good failure data in order to consider it a Type A. Otherwise; the item must be considered Type B.

Our failure data for each element, then allows us to calculate the SFF. The item type and SFF are tabulated below each element in Figure 52.

The redundancy refers to the level of fault tolerance for each element. The level switches operating in a 1oo2 configuration have a fault tolerance, or redundancy of 1. All other items have no fault tolerance and therefore have a redundancy of 0.

Finally, the SIL that can be claimed for the architectural performance of each element can be determined using this information in Table 17.

FIGURE 52. EXAMPLE SAFETY INSTRUMENTED FUNCTION

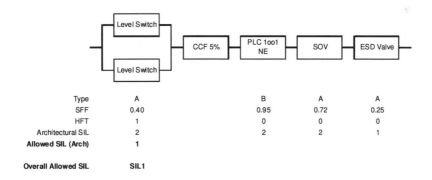

Type	A	B	A	A
SFF	0.40	0.95	0.72	0.25
HFT	1	0	0	0
Architectural SIL	2	2	2	1
Allowed SIL (Arch)	1			
Overall Allowed SIL	SIL1			

The level transmitters are type A, therefore the type A criteria applies. With a SFF of 0.74 and a fault tolerance of 1, the level transmitters comply with the architectural constraints of SIL3.

Similarly, the SOV and ESD Valve can also be assessed. The SOV, also type A, has a fault tolerance of 0 and a SFF of 0.72 giving SIL2. The ESD Valve, type A, fault tolerance of 0 and a SFF of 0.25 giving SIL1.

FIGURE 53. LEVEL SWITCH ARCHITECTURAL CONSTRAINTS

Type A subsystems definition:
Failure modes of all constituent parts well defined, and
Behaviour under fault conditions completely determined, and
Sufficient dependable data from field experience to show that the claimed failure rates for detected and undetected dangerous failures are met.

Safe Failure Fraction	Hardware Fault Tolerance (N)		
(SFF)	0	1	2
<60%	SIL1	SIL2	SIL3
60% - <90%	SIL2	SIL3	SIL4
90% - <99%	SIL3	SIL4	SIL4
>99%	SIL3	SIL4	SIL4

The PLC has been considered a Type B device. This is true for most PLCs because as they are software controlled, there is an element of uncertainty about their failure behaviour and consequently, all of the conditions required for Type A are not met. The assessment of the PLC must therefore be against the Type B requirements.

FIGURE 54. PLC ARCHITECTURAL CONSTRAINTS

Type B subsystems definition:
Failure modes of at least one constituent part not well defined, or
Behaviour of the subsystem under fault conditions cannot be completely determined, or
There is insufficient dependable data from field experience to support the claimed failure rates for detected and undetected dangerous failures.

Safe Failure Fraction	Hardware Fault Tolerance (N)		
(SFF)	0	1	0
<60%	Not allowed	SIL1	SIL2
60% - <90%	SIL1	SIL2	SIL3
90% - <99%	SIL2	SIL3	SIL4
>99%	SIL3	SIL4	SIL4

To summarise, the architectural SIL performance for each element is shown in Figure 52 and the SIL that can be claimed for the whole safety function, is SIL1. The architectural SIL performance for the whole safety function is limited by the lowest SIL claimed.

Requirements for the selection of components and subsystems, IEC61511-1, 11.5

Approach

For Process Sector applications, the selection of components and subsystems can be based on an assessment of suitability. The objectives are to specify requirements:

- for the selection of components and subsystems;

- to enable a component or subsystem to be integrated into the architecture of a SIF;

- to specify acceptance criteria for components and subsystems.

General Requirements, IEC61511-1, 11.5.2

This procedure should not be used for SIL4 applications but for all other components and subsystems the following should be addressed.

The demonstration of suitability must include a SIL Assessment consisting of the calculation of PFD and architectural constraints against the targets.

Demonstration of suitability shall also include consideration of manufacturers' hardware and embedded software documentation. In practice, the documentation accompanying selected components and subsystems will be in the form of technical specifications covering functionality and environmental performance. The FDS must therefore include a statement justifying the suitability of the selected components and subsystems based on the specification documentation available from the manufacturer against the functional requirements.

The components and subsystems must be consistent with the Safety Requirements Specification. In practice, components and subsystems are selected based on their ability to achieve the safety requirements. Demonstration of compliance is by assessment and the requirements for architectural constraints and PFD still apply.

Prior Use, IEC61511-1, 11.5.3

Primarily, component selection should be by procurement specification from approved suppliers. Consideration of the manufacturer's QMS and configuration management systems should be part of vendor assessment and should form part of the evidence of suitability presented in the FDS.

For all selected components and subsystems, the FDS should also reference evidence of accumulated usage. Evidence can be based on either:

- accumulated device hours for SIL1 and field devices;
- accumulated device hours with the identification of dangerous failures for SIL2 and complex items.

For SIL3 logic solver applications, certification is required.

The required accumulated usage for a component or subsystem will depend upon the target failure rate and whether there have been any failures reported. Figure 55 is provided for guidance only and shows the required number of accumulated device years (number of devices x years in use) for various values of target failure rate.

FIGURE 55. GUIDANCE ON REQUIRED USAGE

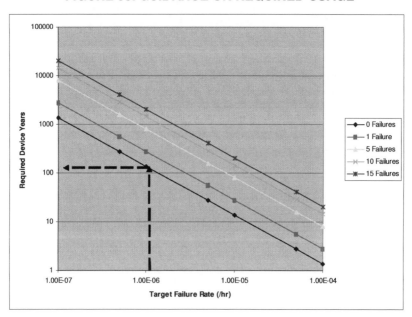

For example, if the target failure rate is 1.00E-06 /hr and zero failures have been reported, then from Figure 55, approximately 137 device years must be evidenced, which can be achieved with 14 devices operating without failure for 10 years. If failures in the field population are reported, then the actual device failure rate will be higher and consequently, more failure free operating hours will be required to demonstrate the same target failure rate.

The figure is based on a X^2-distribution at a 70% confidence limit, and should only be used for guidance and to gain an indication of when a sufficient number of device years have been accumulated.

IEC61511 also requires documented monitoring of returns data and a modification process, by the manufacturer, that evaluates the impact of reported failures.

In practice, failure information is rarely available and selection may therefore include an assessment of the components and subsystems to ensure they will perform as required. This assessment may require discussions with other users, or with manufacturers or users of similar devices or applications. Such supporting evidence should be documented in the FDS as part of the suitability of the components and subsystems.

Fixed Programme Language (FPL) Programmable Devices, IEC61511-1, 11.5.4

Where FPL programmable components and subsystems (for example, field devices) are to be used, the General Requirements, the Requirements for Prior Use and the following requirements for FPL programmable components and subsystems should all be met for SIL1 and SIL2 applications.

In addition, for each selected component, the FDS must justify the selection of FPL components by stating that the component meets the specified requirements in terms of functionality including:

a) characteristics of input and output signals;

b) modes of use;

c) functions and configurations used;

d) unused features are unlikely to impact safety functions.

For SIL3 applications, a formal assessment is required to be carried out.

An alternative approach adopted by some systems integrators is to procure a SIL3 suitable FPL device. These devices should have already undergone a formal assessment by an appropriate organization and SIL3 certification, along with supporting documentary evidence should be provided.

The evidence should show that the device is able to perform the required function and that there is a sufficiently low probability of dangerous failure as a result of random hardware failures, or systematic hardware or software failures. A safety manual should also be available for the device which details operation and maintenance constraints.

Limited Variability Language (LVL) Programmable Devices, IEC61511-1, 11.5.5

Where LVL programmable components and subsystems (for example, logic solvers) are to be used, the General Requirements, the Requirements for Prior Use, the requirements for FPL programmable devices and the following requirements for LVL programmable components and subsystems should all be met for SIL1 and SIL2 applications.

The documentation should present justification that where there is a difference between the operational profile and physical environment as previously experienced, and the operational profile and physical environment when used in the safety function, then the FDS must identify these differences and justify that the PFD will not be adversely affected.

For SIL 1 or 2 applications, a safety configured Programmable Electronic (PE) logic solver (which is a general purpose industrial grade PE logic solver, is specifically configured for use in safety applications) may be used provided that it is justified in the documentation.

The specification documentation available from the manufacturer must show that adequate information covering hardware and software is available to ensure the failure behaviour is understood. This should be confirmed in the FDS by listing all dangerous failure modes and by identifying, where appropriate, diagnostic measures and protective actions. The FDS should also identify the means of protection employed against unauthorised or unintended modification.

For SIL2 logic solver applications, the FDS should confirm the protection technique employed against the following faults during programme execution:

a) program sequence monitoring;

b) protection of code against modifications or failure detection by on-line monitoring;

c) failure assertion or diverse programming;

d) range checking of variables or plausibility checking of values;

e) modular approach;

f) appropriate coding standards have been used for the embedded software.

In addition, the following must be demonstrated:

a) it has been tested in typical configurations, with test cases representative of the intended operational profiles;

b) trusted verified software modules and components have been used;

c) the system has undergone dynamic analysis and testing;

d) the system does not use artificial intelligence nor dynamic reconfiguration;

e) documented fault-insertion testing has been performed.

For SIL2 applications, the FDS should identify constraints for operation, maintenance and fault detection covering the configurations of the PE logic solver and the intended operational profiles.

For SIL3 applications, the documentation should present SIL Certification for any LVL logic solvers.

Full Variability Language (FVL) Programmable Devices, IEC61511-1, 11.5.6

The documentation should present SIL Certification for any FVL logic solvers.

Field Devices, IEC61511-1, 11.6

For the selection of Field Devices, the General Requirements, the Requirements for Prior Use and the following requirements for Field Devices should all be met.

If appropriate, the requirements for FPL programmable devices should also be met.

Field devices shall be selected and installed to minimize failures that could result in inaccurate information due to conditions arising from the process and environmental conditions. Conditions that should be considered include corrosion, freezing of materials in pipes, suspended solids, polymerization, cooking, temperature and pressure extremes, condensation in dry-leg impulse lines, and insufficient condensation in wet-leg impulse lines.

For field devices, the specification documentation should show that the component meets the specified requirements in terms of functionality for all process and environmental conditions and the FDS should confirm this to be the case.

The FDS should also confirm that all energize-to-trip discrete input/output circuits shall apply a method to ensure circuit and power supply integrity, e.g. line monitoring.

Smart sensors must be write-protected to prevent inadvertent modification from a remote location, unless an appropriate safety review allows the use of read/write.

Operator, Maintainer and Communication Interfaces, IEC61511-1, 11.7

For all communication interfaces, the following requirements should be met.

The design of the SIS communication interface must ensure that any failure of the communication interface shall not adversely affect the ability of the SIS to bring the process to a safe state. This should be confirmed in the design documentation.

The documentation should also confirm:

a) the predicted error rate of the communications network;

b) that communication with the BPCS and peripherals will have no impact on the SIF;

c) that the communication interface is sufficiently robust to withstand electromagnetic interference including power surges without causing a dangerous failure of the SIF;

d) the communication interface is suitable for communication between devices referenced to different electrical ground potentials. NOTE: an alternate medium (for example, fibre optics may be required).

Maintenance or Testing Design Requirements, IEC61511-1, 11.8

The SIS design should be such that testing can be carried out either end-to-end or in parts. This takes into account the following as appropriate:

- On-line proof testing – the test design has to ensure that undetected failures can be adequately revealed;

- Test and bypass facilities – an operator should be alerted if any portion of the SIS is bypassed for maintenance or testing purposes;

- Forcing of inputs and outputs without taking the SIS offline should not be allowed unless adequate procedures and security are in place. As per the bypass function, an operator needs to be alerted if any inputs or outputs are forced.

SIF Probability of Failure, IEC61511-1, 11.9

Refer to Section [14].

Requirements for Application Software, IEC61511-1, 12

IEC61511-1, 12 lists the requirements that apply to any software forming part of a SIS, or used to develop a SIS. The requirement defines the application software safety life-cycle requirements to ensure that:

- all activities required to develop the application software are defined;

- the software tools that are used to develop and verify the application software, i.e. the utility software, is fully defined;

- a plan is put in place to meet the functional safety objectives.

The general requirement is to define the applicable phases of the software safety life-cycle to be considered and document all relevant information. These include the following:

- software safety requirements specification – similar to the hardware requirements, a specification needs to be defined which lists all the software safety requirements in a clear and structured manner which enables the design team to develop the application software accordingly;

- software safety validation planning – this should be carried out as part of overall SIS validation planning;

- design and development – the application software needs to be developed to meet the system design requirements, outlined in the software SRS, in terms of safety functions and safety integrity levels. Suitable languages, programming and support tools which assist verification, validation, assessment and modification should be used. The design should be modular and structured, in a way that achieves testability and allows safe modification. Adequate software module testing should be carried out to verify the correct functionality. Note that verification should be carried out for each phase of the software safety life-cycle;

- integration – once tested and verified, the software needs to be integrated with the SIS subsystem and tested in order to demonstrate that that it meets the requirements in the SRS when running on the hardware;

- software safety validation – this should be carried out as part of the overall SIS validation (phase 5);

- modification - any modification of validated software should be carried out in a controlled manner such that the software integrity is maintained.

14. SIF PROBABILITY OF FAILURE, IEC61511-1, 11.9

Complying with the Standard

So far we have identified that we must establish target reliability measures in order to ensure that the overall risk does not exceed the maximum tolerable risk.

We have also seen that the target reliability measure can be expressed in SILs and to comply with the standard, we not only have to demonstrate that the safety function meets quantitative targets but also that we apply appropriate controls.

The reason why the target reliability measures are banded into SILs is because the standard demands that the management of functional safety at various parts of the lifecycle, become more rigorous, the higher the SIL.

Complying with the standard requires that the target reliability measures are achieved appropriate to the SIL applied.

SIL Target Reliability Requirements

The PFD for each SIL depends upon the mode of operation in which a SIS is intended to be used, with respect to the frequency of demands made upon it. These are defined in section [6] and may be either:

Demand Mode, where a specified action is taken in response to process conditions or other demands. In the event of a dangerous failure of the SIF a potential hazard only occurs in the event of a failure of the process of BPCS;

Continuous Mode, where in the event of a dangerous failure of the SIF a potential hazard will occur without further failure unless action is taken to prevent it.

Based on these criteria, the appropriate targets presented in Table 18 can be applied.

TABLE 18 SIL SPECIFIED PFD AND FAILURE RATES

SIL Level	DEMAND MODE Probability of failure on demand	HIGH DEMAND Failure rate per hour
SIL 4	$\geq 10^{-5}$ to $< 10^{-4}$	$\geq 10^{-9}$ to $< 10^{-8}$
SIL 3	$\geq 10^{-4}$ to $< 10^{-3}$	$\geq 10^{-8}$ to $< 10^{-7}$
SIL 2	$\geq 10^{-3}$ to $< 10^{-2}$	$\geq 10^{-7}$ to $< 10^{-6}$
SIL 1	$\geq 10^{-2}$ to $< 10^{-1}$	$\geq 10^{-6}$ to $< 10^{-5}$

Calculation of PFD for a Demand Mode Safety Function

When performing reliability calculations, we assume that failures occur randomly in time with a constant rate and when a failure occurs, the failed element will be unavailable until the failure is detected and repaired.

In the calculation of PFD, we are essentially calculating the probability that the SIS will be unavailable when a demand is placed upon it. For a 1oo2 redundant system, given that we have had one channel failure, the PFD is the probability that the second channel subsequently fails during the down time of the first.

The following general relationships can be used in the calculation of PFD. The equations used are simplifications to the standard equations and are derived in [20.8].

For detected failures: Ref. IEC61508-6, B.3.2.2.1

$$PFD_{1oo1} = \lambda_{DD} . MDT$$

$$PFD_{1oo2} = \lambda_{DD}^2 . MDT^2 + \beta . \lambda_{DD} . MDT$$

Ref. IEC61508-6, B.3.2.2.2

For undetected failures: Ref. IEC61508-6, B.3.2.2.1

$$PFD_{1oo1} = \lambda_{DU} . T_p / 2$$

$$PFD_{1oo2} = \lambda_{DU}^2 . T_p^2 / 3 + \beta . \lambda_{DU} . T_p / 2$$

Ref. IEC61508-6, B.3.2.2.2

Where λ_{DD} is the dangerous detected failure rate, λ_{DU} is the dangerous undetected failure rate and β is the contribution from common cause failures. T_p is the proof test interval and MDT is the Mean Down Time.

Failure Rates

In the calculation of PFD and SFF, the analysis uses the underlying hypothesis of IEC61508-6, Annex B.3 in that component failure rates are constant over the lifetime of the system.

The failure rates used in calculations may be obtained by FMECA, quantified by field data, or by reference to published data from industry sources. The failure rates used should be compared with available data for similar modules of complexity and technology. This approach ensures a conservative approach in terms of reliability modelling and gives confidence that the calculated reliability performance should be achievable in service.

Reliability Modelling

In this example, the process and SIF are highlighted, Figure 56.

FIGURE 56. DEMAND MODE SAFETY INSTRUMENTED FUNCTION

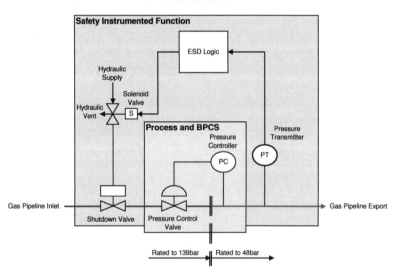

The calculation of PFD is most easily performed using the Reliability Block Diagram (RBD) technique. In RBDs, the diagrams show the items or components required for a reliable system and do not necessarily represent physical layout or connections. RBD modelling is a valid technique described in IEC61508-6, Annex B, 4.2.

The RBD for the SIS described is shown, Figure 57.

FIGURE 57. RBD FOR DEMAND MODE SAFETY FUNCTION

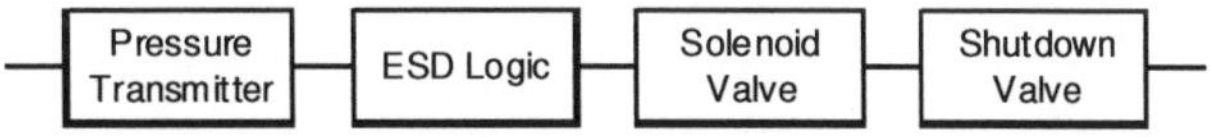

	Pressure Transmitter	ESD Logic	Solenoid Valve	Shutdown Valve
λDD	2.64E-07	3.42E-06	0.00E+00	0.00E+00
MDT	48	48	48	48
Configuration PFD	1.27E-05	1.64E-04	0.00E+00	0.00E+00
λDU	4.00E-08	1.63E-07	6.00E-07	4.64E-06
Proof Test Period	8760	8760	8760	8760
Configuration PFD	1.75E-04	7.14E-04	2.63E-03	2.03E-02

PFD (revealed)	1.77E-04
PFD (unrevealed)	2.38E-02
PFD	**2.40E-02**
Allowed SIL (PFD)	**SIL1**

The RBD shows the calculation of PFD. Below each element are the values for λ_{DD} dangerous detected failure rate, λ_{DU} dangerous undetected failure rate, MDT, the mean down time and proof test period T.

Example Demand Mode Prepolymer Loop Modification Safety Integrity Level Assessment

The following describes an example of a SIL Assessment of the PFD and architectural performance of a SIF.

Scope

The ESD function, S-005 prevents a runaway reaction in 39-R-050, and consequently protects against loss of containment from the reactor which could result in operator injuries and also subsequent environmental damage. Currently, safety function S-005 is initiated by detection of high temperature or high pressure in the reactor and relief valve ROV0503 is opened to relieve the pressure.

It is understood that there were concerns that ROV0503 may not provide sufficient capacity for the relief case and therefore the ESD action of S-005 has been modified to include the activation of an additional relief valve ROV0501. In addition, during the upgrade programme, two hand switches (Permissive HS0900 and Override HS2004) have been included for maintenance purposes.

Objectives

The client maintains a large number of sensors as part of the SIS, and is keen to minimise this overhead. The objective of this analysis is therefore to:

a) determine which items should be included in an analysis of the modified ESD safety function S-005;

b) construct a RBD to determine the PFD and architecture of S-005;

c) suggest a proof test philosophy (test intervals for sensors, handswitches, logic and relief valves) that allows the targets (Table 19); to be met whilst minimising the sensor test frequency.

Note: the client has stated that proof test intervals for any item should not exceed 36 months. From an engineering point of view, the client is not comfortable with parts of the SIS not being exercised for long periods of time.

Permissives and Overrides

There are two hand switches, HS2004 and HS0900 that are associated with ESD S-005.

It is understood that HS0900 is used to direct the catalyst to the reactor and consequently, if the switch is in the wrong position, or fails in the wrong state, the hazard cannot occur. HS2004 is used as a trip override on S-005. If HS2004 is inadvertently left in the override position following maintenance action, or if it fails in the override state, then the safety function S-005 will be disabled.

Hardware Configuration

The logic solver is based on a Triple Modular Redundant (TMR) configuration voted 2 out of 3 (2oo3).

Figure 58 presents a schematic of the configuration of the hardware.

FIGURE 58 HARDWARE SCHEMATIC

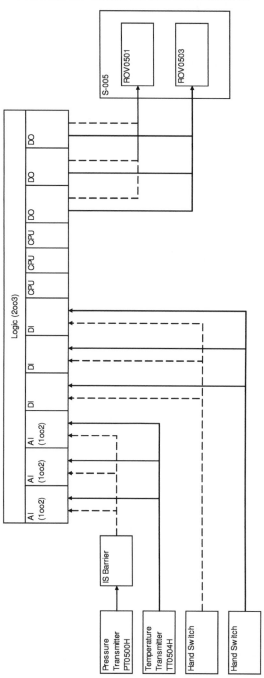

Safety Functions Analysed

Table 19 presents the established SIL and PFD targets.

TABLE 19 SAFETY FUNCTIONS FOR ANALYSIS

Loop	Initiator	ESD Action	Conditions required to mitigate the hazard	PFD Target	SIL Target
1	High pressure [PT0500H] or high temperature [TT0504HH]	Activates S-005	ROV0503 and ROV0501 open.	5.56E-03	SIL2

Diagnostic Coverage

It has been assumed that all undetected failure modes will be revealed by the proof test, i.e. by full execution of the SIS function.

Mean Down Time

The MDT of 72 hours should be used in this analysis.

Proof Test Period

The proof test intervals should be selected to achieve the targets whilst maximising the sensors test interval.

Accounting for Common Cause Failures

CCFs are failures that may result from a single cause but simultaneously affect more than one channel. They may result from a systematic fault for example, a design specification error or an external stress such as an excessive temperature that could lead to component failure in both redundant channels.

The contribution of CCFs in parallel redundant paths should be accounted for in the model, by inclusion of a β factor. The CCF failure rate that is included in the calculation is equal to β x the total failure rate of one of the redundant paths. The β factors to be used in the analysis are summarised in Table 20.

TABLE 20. β FACTORS

Redundant Configuration	β Factor	Justification
Sensors PT0500, TT0504	3%	Since the sensors are different technology measuring different process variables the potential for common cause failures is limited to the process itself, the mechanism for attaching the sensors and the routing and separation of sensor connections. The value of 3% is therefore judged to be reasonably conservative.
PLC TMR Logic	5%	Common cause failures in a redundant TMR configuration are small however, a value of 5% has been used in order to maintain a conservative approach.

Type A Components

The following items may be considered to be Type A:

- IS Barrier (Transmitter Power Supply Isolator; PB0500);
- Temperature Transmitter (TT0504);
- Hand switch (HS0900, HS2004)
- Pre-polymerisation relief valves.

Type B Components

The following items were considered to be Type B:

- PLC Logic Modules;
- Pressure transmitters (PT0500).

Failure Rates of Components

The analysis should assume constant failure rates since the effects of early failures are expected to be removed by appropriate processes. These processes include the use of mature products from approved sources, in-house testing prior to delivery and extended operation and functional testing as part of installation and commissioning. Field returns data on other similar projects indicates that early life failures do not result in a significant number of returns and therefore the techniques employed are judged to be sufficient.

It is also assumed that components are not operated beyond their useful life thus ensuring that failures due to wear-out mechanisms do not occur.

The failure rates (in failures/hour) that may be used in the model in the calculation of PFD, λ_{DD} and λ_{DU}, are summarised in Table 21. The failure rates were obtained from a combination of sources.

TABLE 21. FAILURE RATES (/HR) AND THE CALCULATION OF SFF

Item Ref/ Tag Description	λ	λD	λDU	λDD	λS	SFF
Input devices						
PT 0500 Pressure Transmitter (IS)	1.5E-06	1.4E-06	6.0E-07	7.5E-07	1.5E-07	0.60
PT 0501 Pressure Transmitter (IS)	1.5E-06	1.4E-06	6.0E-07	7.5E-07	1.5E-07	0.60
PB 0500 Barrier - for PT above (non IS)	2.1E-07	6.3E-08	6.3E-08	0.0E+00	1.5E-07	0.70
PB 0501 Barrier - for PT above (non IS)	2.1E-07	6.3E-08	6.3E-08	0.0E+00	1.5E-07	0.70
FT 0041 Coriolis Flowmeter	2.6E-06	2.2E-06	9.0E-07	1.3E-06	4.0E-07	0.65
TT 0504 3 wire RTD with head mounted transmitter	2.0E-06	1.4E-06	4.0E-07	1.0E-06	6.0E-07	0.80
HS 2004 Override Switch	2.0E-06	8.0E-07	8.00E-07	0.0E+00	1.2E-06	0.60

Item Ref/ Tag Description	λ	λD	λDU	λDD	λS	SFF
HS 0900 Permissive Switch	2.0E-06	8.0E-07	8.0E-07	0.0E+00	1.2E-06	0.60
Logic devices						
CPU	1.5E-06	5.2E-07	6.4E-09	5.1E-07	9.9E-07	1.00
32pt DI Module	2.2E-08	1.1E-08	9.9E-11	1.1E-08	1.qE-08	0.99
32pt AI Module	1.4E-08	7.0E-09	9.9E-11	6.9E-09	7.0E-09	0.99
16pt DO Module	2.9E-08	1.5E-08	9.9E-11	1.5E-08	1.5E-08	0.99
Output devices						
39-PM-050 Pump running status from contactor & relay NO Contact	3.0E-07	2.0E-07	1.9E-07	0.0E+00	1.1E-07	0.35
ROV 0501 AOV (FO) Dump Valve including SOV	5.1E-06	1.4E-06	1.4E-06	0.0E+00	3.7E-06	0.73
ROV 0503 AOV (FO) Dump Valve including SOV	5.1E-06	1.4E-06	1.4E-06	0.0E+00	3.7E-06	0.73
ROV 0404 AOV (FC) including SOV	9.7E-06	3.0E-06	3.0E-06	0.0E+00	6.7E-06	0.68
ROV 0405 AOV (FO) Dump Valve including SOV	5.1E-06	1.4E-06	1.4E-06	0.0E+00	3.7E-06	0.73
ROV 0406 AOV (FO) Dump Valve including SOV	5.1E-06	1.4E-06	1.4E-06	0.0E+00	3.7E-06	0.73

<u>One Possible Solution</u>

The RBD shown in Figure 59 shows the elements required as part of the safety function. It is not necessary to include HS0900 in the safety function assessment as its failure cannot prevent the safety function from operating. If HS0900 fails or is left in the wrong position, then the hazard cannot occur.

HS2004 must be included because if it is inadvertently left in the override position following maintenance action, or if it fails in the override state, then the safety function S-005 will be disabled.

The calculation of PFD required some judgement to be applied with regard to the setting of proof test intervals, Tp. The requirement was to maximise the interval up to 3 years whilst achieving the target PFD. There will be many potential solutions and in practice this would be discussed with the client. A possible proof test philosophy is shown in Table 22.

TABLE 22. POSSIBLE PROOF TEST INTERVALS

Proof Test Period (sensors)	24	mths	17520	hrs
Proof Test Period (hand switch)	6	mths	4380	hrs
Proof Test Period (logic)	36	mths	26280	hrs
Proof Test Period (valves)	3	mths	2190	hrs

These proof test intervals provide a calculated PFD of 4.91E-03 against a target of 5.56E-03 and the PFD and architectural performance both meet the target SIL2.

A possible solution RBD is presented in Figure 59.

FIGURE 59 SOLUTION RBD

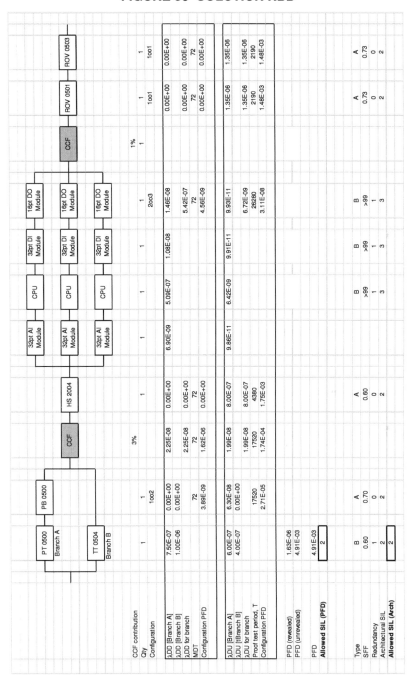

Traceability of Failure Rate Data

In performing PFD calculations it is vital that all calculations are visible and all data used is traceable to source. Microsoft excel is a useful tool as it fulfils both of these requirements and it also allows a graphical representation of the reliability model to be developed, as in Figure 59.

The spreadsheet allows each data cell to point to a data table where all the collected failure rate data and data sources can be presented. An example data table is shown in Figure 23. It is important that the data source reference is detailed enough so that anyone can check and confirm the values used.

If using an excel format, it is convenient to also list the component type and assumed MDT and Tp used in the calculation. This allows the proof test interval to be changed easily and the effect on PFD to be automatically calculated.

TABLE 23. TYPICAL DATA TABLE

Item / Part Number	$\blacklozenge$	$\blacklozenge$D	$\blacklozenge$DD	$\blacklozenge$DU	$\blacklozenge$S	Type	SFF	MDT	Tp	Data Source
PT0500	1.35E-06	8.18E-07	7.50E-07	6.80E-08	5.27E-07	B	0.95	4380	4380	exida
SIL3 Logic Soilver	5.57E-06	2.23E-06	2.21E-06	2.20E-08	3.34E-06	B	1	168	4380	Sintef
Analogue Input Module	1.07E-06	5.34E-07	5.08E-07	2.60E-08	5.34E-07	B	0.98	168	4380	Sintef
Discrete Output Module	5.26E-07	2.63E-07	2.50E-07	1.30E-08	2.63E-07	B	0.98	168	4380	Sintef
12" HIPPS Valve.	5.29E-06	2.12E-06	0.00E+00	2.12E-06	3.17E-06	A	0.6	730	4380	Oreda 2002

Sources of Failure Rate Data

Failure rate data should only be obtained from appropriate sources and this will depend upon the application. The following lists data sources that have been used, and may be appropriate in the process sector.

Exida.com Safety Equipment Reliability Handbook, 2007, 3rd Edition

Volume 1 – Sensors, ISBN 978-0-9727234-3-5

Volume 2 – Logic Solvers and Interface Modules, ISBN 978-0-9727234-4-2

Volume 3 – Final Elements, ISBN 978-0-9727234-5-9.

Handbook of Reliability Data for Electronic Components used in Telecommunications Systems, HRD-5.

Hydrocarbon Leak and Ignition Database Report No. 11.4/180 May 1992.

IEEE Standard 500-1984. Guide to the Collection and Presentation of Electrical, Electronic, Sensing Component, and Mechanical Equipment Reliability Data.

OREDA, The Offshore Reliability Data Handbook 4th Edition 2002 ISBN 82-14-02705-5.

Parloc 2001: 5th Edition, The Institute of Petroleum, published by the Energy Institute ISBN 0 85293 404 1.

Reliability Data for Control and Safety Systems, 2006 Edition, PDS Data Handbook, SINTEF, ISBN 82-14-03898-7.

Reliability Technology, AE Green and AJ Bourne, Wiley, ISBN 0-471-32480-9.

15. DESIGN AND ENGINEERING OF THE SIS

Lifecycle Phases

Figure 38 shows the phase of the lifecycle that applies.

FIGURE 60. LIFECYCLE PHASE 4

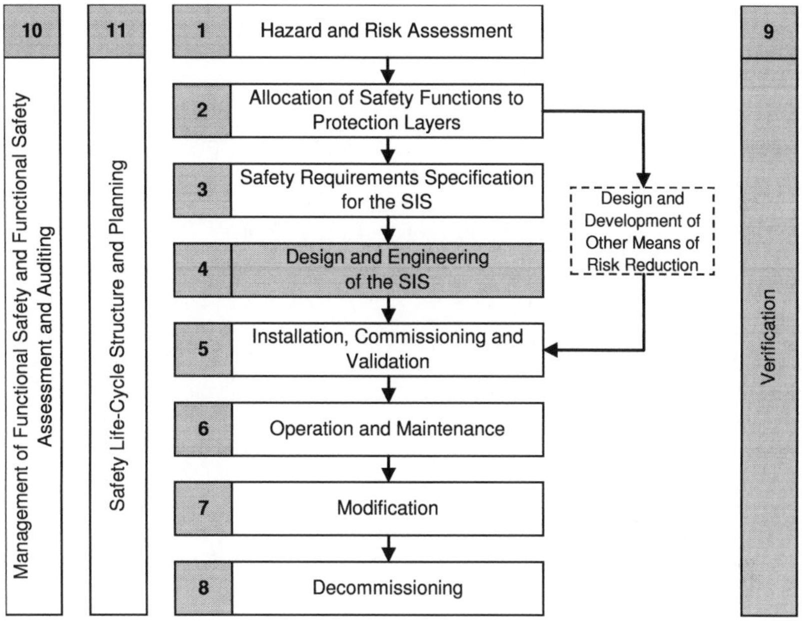

The objective of Lifecycle Phase 4 as defined in IEC61511-1, 11.1 is to:

- Design the SIS in order to provide the necessary SIFs;
- Verify that the SIF design meets the specified SIL, defined during the SIL determination [13].

SIF Design

The SRS will form the basis of the SIF design and will enable the design team to translate the functionality into design documents such as a FDS. Thus, the FDS should contain all the functional and integrity requirements that are needed to design and engineer the SIS.

It is important that the design documentation includes the following requirements:

- Requirements for system behaviour on detection of a fault;

- Hardware fault tolerance;

- Selection of components and subsystems;

- Field devices;

- Operator, maintainer and communication interfaces with the SIS;

- Maintenance or testing design requirements;

- SIF probability of failure;

- Application software.

These requirements are examined in more detail in [13]. The following additional guidance on SIF design is provided.

Requirements for system behaviour on detection of a fault

The requirements for the behaviour of the SIF, on detection of a fault should be understood and documented in the SRS.

Failure behaviour should consider the effects of power failure or loss of utilities such as hydraulic power or instrument air. For ESD systems, loss of power should result in the safe state. ESD Systems are normally energised, and de-energise to trip to the safe state. Therefore all power faults should be safe failures.

F&G Systems are normally de-energised and alarms, beacons and deluge systems must be energised to be activated. In this respect there is no clear safe state as both the de-energised and the energised states can be considered safe. As such, the desired state on power failure should be specified in the SRS and the SIF should be engineered to achieve this.

For F&G Systems, and other systems that do not cause a trip on loss of power, all of the following should be implemented:

- line monitoring to detect failed output;

- supplemental power using battery backed uninterruptible power supplies;

- detection of loss of power to the subsystem.

Where a SIF is configured as a redundant channel, behaviour on channel failure should be specified in the SRS. For example, where a redundant configuration of sensors is voted by the logic solver, then the logic voting should adapt on detection of a channel fault.

If one pressure sensor fails to the trip condition (safe) in a 2oo3 configuration, then in this example, only one additional trip signal is required from either of the two remaining sensors to initiate executive action. The voting has effectively adapted from 2oo3 to 1oo2 on detection of a (dangerous) failure.

If the pressure sensor fails to a safe state and the failure is detected, the logic solver could set the input to the fault condition and adapt the voting from 2oo3 to 1oo2.

If a fault is detected in a 1oo2 configuration, then the logic could set the input from the sensor to the fault condition and adapt the voting to trip on the remaining sensor. The voting has effectively adapted from 1oo2 to 2oo2 on detection of a (dangerous) failure.

Hardware fault tolerance

The architecture of the SIF will depend upon the SIL target of the function and also the SFF of the function subsystems in accordance with Table 17.

In designing the SIF therefore, the SFF of all components must be determined and understood. SFF can sometimes be obtained from published data or assessments that manufacturers have carried out or commissioned. Occasionally it may be necessary to conduct a Failure Modes, Effects and Criticality Analysis (FMECA) on an item to determine and quantify the failure effects to calculate SFF.

There are other factors to consider in determining the architecture of a SIF. A 1oo2 architecture has the same HFT as a 2oo3, or a 3oo4 architecture, in that only one fault can be tolerated. All of these configurations therefore may meet the HFT requirements of the SIL but each will result in a different PFD. Each configuration will also result in a different STR, and whilst there are no STR targets specified in the standard, spurious trips have commercial implications.

The following figure gives an illustration of the relationship between calculated values of PFD against STR for common configurations of a typical safety function. Note that the actual values of PFD and STR will differ for each SIF design, with the actual devices, the failure rates and test intervals used. Figure 61 gives a general indication.

FIGURE 61. PFD AND STR FOR COMMON CONFIGURATIONS

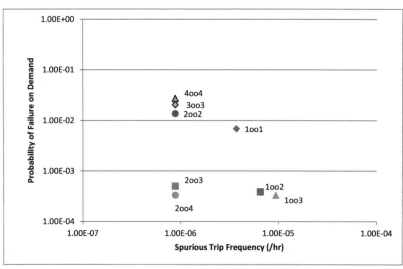

The figure shows that simplex (HFT = 0) can typically be expected to fall within the SIL1 band, with 1oo1 possibly achieving SIL2 and redundant configurations being required for SIL3.

If SIL3 is the target, then 2oo3 and 2oo4 generally have a lower STR than 1oo2 and 1oo3 but all may meet the PFD targets. The additional capital and scheduled maintenance costs of the 2oo3 and 2oo3 must be weighed up against the cost of the higher spurious trip rate

Note that the purpose of the figure is to give an indication of the relative positions of each configuration in terms of the trade-off between PFD and STR. In designing the configuration therefore, the HFT and PFD targets are primary considerations, and must be achieved. The STR may also be addressed as a secondary commercial consideration.

Selection of components and subsystems

In selecting components and subsystems for use in SIFs, it is a requirement to demonstrate that the items selected are suitable in order to give confidence that the required safety integrity is achieved in practice.

There are specific suitability requirements that apply to: field devices, operator, maintainer and communication interfaces and application software. Options available for demonstrating suitability are described earlier [13] and it is the duty of the SIF designer to collect appropriate evidence which may include usage data in appropriate applications and environments, or third-party certification or assessment.

Maintenance or testing design requirements

The SIF should be designed so that proof tests can be carried out. It is important that the SIF addresses the hazard and not the causes of the hazard or the consequential effects. For example, if impurities in a reactor can lead to over temperature and high pressure which results in vessel rupture, the SIF should monitor pressure and trip on high pressure, not temperature.

To Proof Test a pressure transmitter that initiates a trip on high pressure, it should be presented with a pressure environment which is increased until the trip level is exceeded and the trip signal is output. The test pressure should be presented to the transmitter sensing element, as it would be in normal operation. It is not sufficient to inject an electronic signal downstream of the sensing element to simulate a pressure reading.

When proof testing redundant configurations, it is necessary to check that all redundant paths behave as specified. For example, if the executive action is to close two ESDVs configured as 1oo2, then the proof test should verify that both valves have closed successful even though only one is required to close to achieve the safe state.

Whether tight shut-off verification is additionally required will depend upon whether or not tight shut-off is necessary to achieve the safe state.

It is permissible to proof test sections of the SIF separately, i.e. sensor subsystem, logic, or final element, provided the test coverage overlaps each subsystem and no part of the SIF remains untested. Similarly, the proof testing of subsystems can be carried out at different test intervals provided this is reflected in the calculation of PFD and the targets are achieved.

SIF probability of failure

Care should be taken in deciding what should be included as part of the SIF, when calculating PFD. It is common practice for the executive action to trip the final elements necessary for the safe state and also to initiate a number of other actions for good housekeeping reasons, e.g. to ensure shutdown occurs in a controlled manner or to enable subsequent re-start to be achieved more easily. However, when constructing the SIF calculation it is important to only include those elements that are required to function in order to achieve the safe state.

In general, for any configuration, the calculation of PFD is dominated by the contribution of the dangerous undetected failure rates of the SIF components, and the proof test interval, T_P. Typically, T_P is set to 1 year (8760 hours) as this often coincides with the scheduled plant shutdown and it is convenient to proof test all SIFs when the plant is not operating. Proof testing usually trips the plant.

The target T_P will be specified in the SRS and this should be used in the calculation of PFD. In practice, proof test intervals should not exceed either the interval specified in the SRS, or an interval considered to be consistent with good engineering practice, whichever is the shorter. The reason for this is that the PFD target may be achieved with a theoretical proof test interval of 20 years or more, but in practice, this would result in the SIF remaining untested, and the valves and final elements not being exercised for many years. There would be a significant possibility that when a demand was placed on the SIF, some corrosion, contamination or blockage would prevent the trip.

It is worth reiterating that even though compliance with the standard may be demonstrated, this does not exonerate the duty-holder from responsibility in the event of an accident.

Overrides

In designing SIFs, it is sometimes desirable to provide bypass facilities to allow on-line testing, i.e. proof testing without taking the process off line.

Overrides should always set an alarm when in bypass to reduce the likelihood that the bypass is left in place inadvertently.

Maintenance and test facilities should be designed to minimise the likelihood of dangerous failures arising from their use. When SIFs are bypassed, the duty-holder has the responsibility of taking some other mitigating action in order to minimise the risk.

Separation of SIF and BPCS

The standard requires a certain degree of separation between the SIF and the BPCS that it is protecting. The concern is that, if there is some commonality between the two systems, then failure of the BPCS may also cause the SIF to fail.

The design of the SIF, and of any protection layers that have been taken credit for, should therefore take into account common cause, common mode and dependent failures between protection layers, and also between protection layers and the BPCS. The PFD analysis should include contributions from common cause failures.

Diversity between protection layers and the BPCS is desirable but not always achievable. An example could be over-pressure protection of a vessel, where the BPCS and the SIF both require pressure measurement and there will be a limit on the suitable equipment available. Some diversity can be achieved by using equipment from different manufacturers but the diversity may be of limited value.

Wherever possible, SIFs should be separated from non-safety functions in such a way that failure of a non-safety function, or operator access to non-safety software, can not cause a dangerous failure of the SIF. Where the SIS implements both safety and non-safety functions then all of the hardware and software that can affect any SIF under normal or fault conditions should be treated as part of the SIS and comply with the requirements of the highest SIL.

Similarly, sensors or valves for example, which are used by a SIF may also be used by the BPCS provided it can be shown that a failure of the BPCS can not adversely affect the SIS.

Where the SIS provides SIFs with different SIL requirements, then the shared or common hardware and software should meet the highest SIL target.

The SIS shall be designed such that once it has placed the process in a safe state, it shall remain in the safe state until a reset has been initiated unless otherwise directed by the Safety Requirements Specifications.

Partial Stroke Testing

Partial stroking of shutdown valves is permitted and provides a good way of exercising valves and actuators without causing loss of production. Partial stroke testing effectively introduces some fault detection to devices that typically have no diagnostic coverage.

The effectiveness however may depend upon the application. For valves operating in clean service, partially stoking the valve will enable a stuck valve to be diagnosed, and this failure may be the dominant mode in this application. For a valve operating in a dirty inventory, the dominant failure mode may be that the valve fails to close and seat properly due to a build-up of deposits in the valve mechanism. Partial stroke testing in this application may not be significantly beneficial.

In some applications therefore, partial stroking can allow the proof test interval to be extended, but it cannot replace the proof test.

16. INSTALLATION, COMMISSIONING AND VALIDATION, IEC61511-1, 5

Lifecycle Phases

Figure 62 shows the phase of the lifecycle that applies.

FIGURE 62. LIFECYCLE PHASE 5

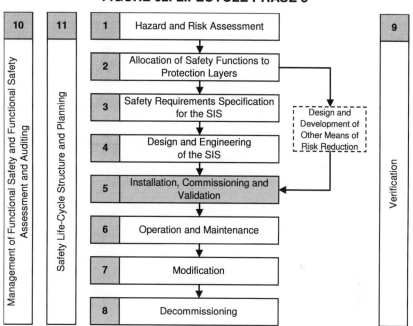

The objectives of Lifecycle Phase 5 defined in IEC61511-1, 14 and 15 are to:

- install the SIS according to the specifications and documentation;

- commission the SIS so that it is ready for final system validation;

- validate that the installed and commissioned SIS achieves the requirements defined in the SRS.

SIF Installation

The requirements for installation should be defined in the Installation and Commissioning Plan or integrated into the overall project plan. Installation procedures should define the activities to be carried out, the techniques and measures to be used, the persons, departments or organisations responsible and the timing of the installation activities.

SIF Commissioning

The SIS should be commissioned in accordance with planning and procedures. Records should be produced stating the test results and whether the acceptance criteria defined during the design phase, have been met. Failures should be investigated and recorded.

Where it is established that the actual installation does not conform to the design information, then the difference should be investigated and the impact on safety determined.

SIF Validation

The validation procedures should include all modes of operation of the process and associated equipment and should include:

- start-up, normal operation, shut-down;
- manual or automatic operation;
- maintenance modes, bypassing constraints;
- timing;
- roles and responsibilities;
- calibration procedures.

In addition, validation of application software should include:

- identification of software for each mode of operation;
- validation procedure to be used;
- tools and equipment to be used;
- acceptance criteria.

The validation should ensure that the SIS performs under all modes of operation and is not affected by interaction of the BPCS and other connected systems. Performance validation should ensure that all redundant channels operate, bypass functions, start-up overrides and manual shutdown systems operate.

The defined, or safe, state should be achieved in the event of loss of energy, e.g. electrical or hydraulic power, or instrument air.

Diagnostic alarm functions defined in the SRS should operate and perform as specified on invalid process variables, e.g. out of range inputs.

Following validation, appropriate records should be produced and identify the test item, test equipment, test documents and test results including any discrepancies and analyses or change requests made as a result.

17. OPERATION AND MAINTENANCE, IEC61511-1, 6

Lifecycle Phases

Figure 63 shows the phase of the lifecycle that applies.

FIGURE 63. LIFECYCLE PHASE 6

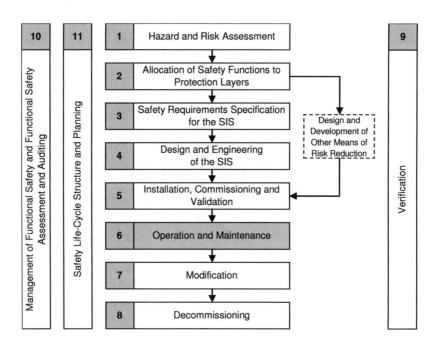

The objectives of Lifecycle Phase 6 as defined in IEC61511-1, 16.1 are to:

- Ensure that the required SIL of each SIF is maintained during operation and maintenance;

- Operate and maintain the SIS so that the designed functional safety is maintained.

SIF Operation and Maintenance (O&M)

The requirements for O&M should be defined in the O&M Plan or integrated into the overall project plan. O&M procedures should define the routine operations that need to be carried out to maintain the functional safety of the SIS.

These routine operations should include requirements for:

- proof testing;
- bypassing a SIF for test or repair;
- routine collection of data: e.g. results of audits and tests on the SIS, records of SIF demands, failures and repair and proof test downtimes.

Proof test procedures should be developed so that every SIF is tested to reveal dangerous failures that remain undetected by diagnostics.

Maintenance procedures are required for fault diagnostics, repair, system revalidation following repair action, actions to be taken following discrepancies between expected behaviour and actual behaviour, calibration and maintenance of test equipment and maintenance reporting.

Reporting procedures are required for reporting failures, analysing systematic and common cause failures and for tracking maintenance performance.

O&M Training

Training of O&M staff should be planned and carried out in good time so that the SIS can be operated and maintained in accordance with the SRS. Training should include:

- hazards;
- trip points;
- executive actions;
- operation of all bypasses and any constraints on their use;

- manual operations, e.g. start-up, shut-down and any constraints on their use;

- operation of alarms and diagnostics available.

Proof Testing

Proof test procedures should test the complete SIF from sensing element to final actuated device. The proof test interval should be that used in the quantification of PFD [14].

It is acceptable to test different elements of the SIF at different intervals, provided:

- the calculated PFD is still acceptable;

- there is some overlap in the test so that no part of the SIF remains untested.

18. MODIFICATION AND DECOMMISSIONING, IEC61511-1, 7, 8

Lifecycle Phases

Figure 64 shows the phases of the lifecycle that apply.

FIGURE 64. LIFECYCLE PHASES 7, 8

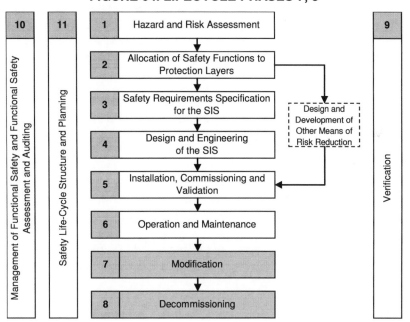

The objectives of Lifecycle Phase 7 and Lifecycle Phase 8 as defined in IEC61511-1, 17.1 and 18.1 are to ensure that:

- any modifications to any SIF are properly planned, reviewed and approved prior to making the change;

- the required safety integrity is maintained following any changes that may be made;

- prior to decommissioning, a proper review is conducted and authorisation obtained to ensure that the safety integrity is maintained during decommissioning.

SIF Modification

Prior to carrying out any modifications, procedures should be in place for authorising and controlling the changes. This will typically be handled with a Change Request Note (CRN) which usually forms part of a QMS.

Any change request should describe the change required and the reasons for the request. This may be raised by O&M staff as a result of incidents during operation or maintenance. The usual approval process for change requests should involve different departments within an organisation to determine the impact of the change on design, the installed base, and the required implementation.

Once an organisation is involved in functional safety, any change request should in addition, be reviewed by a competent person, e.g. the Safety Authority (SA) to determine whether the change can affect safety and if so, an appropriate impact analysis is required.

Impact Analysis

The results of the analysis may require the early parts of the lifecycle to be revisited and for example, it may be necessary to review identified hazards and risk assessments. Modification activities cannot begin until this process has been completed and the SA has authorised the change.

The impact of changes to the SIF may have consequential effects on O&M personnel and additional training may be necessary.

SIF Decommissioning

Decommissioning should be a planned activity as part of lifecycle phase 11, and may be treated as a modification at the end of the project life.

Commencement of the decommissioning phase should initiate an impact analysis to determine the effect of decommissioning on functional safety. The analysis should include revision of the hazard identification and risk assessment with particular consideration of the hazards that may occur as a result of the decommissioning activity.

19. MANAGEMENT OF FUNCTIONAL SAFETY, ASSESSMENT AND AUDITTING

Lifecycle Phases 10 and 11

Figure 65 shows the phase of the lifecycle that applies.

FIGURE 65. LIFECYCLE PHASE 10 AND 11

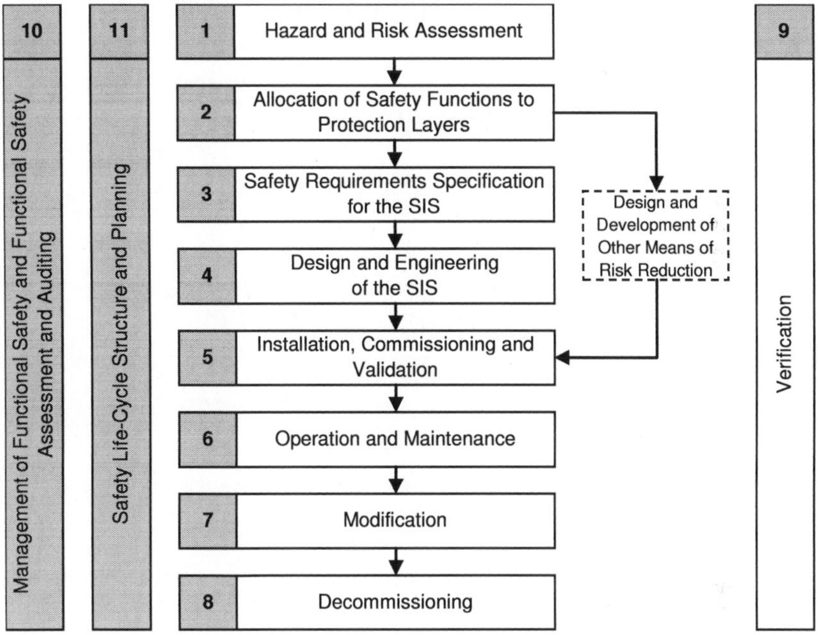

The objective of this phase, as defined in IEC61511-1, 5 is to identify the management activities and documentation necessary to enable the applicable lifecycle phases to be adequately addressed by those responsible.

The standard lists general requirements for management and documentation to enable the applicable lifecycle phases to be adequately addressed by those responsible.

This means that project documentation must contain sufficient information for each phase of the overall lifecycle that has been completed, for the subsequent phases and for the verification activities, to be completed effectively.

Compliance with the standard requires the specification of:

- responsibilities in the management of functional safety;
- the activities to be carried out by those with responsibilities.

Compliance against the requirements can be addressed by putting procedures in place that deal with each requirement in scope, by implementing those procedures and by ensuring there is adequate information available to enable the management of functional safety to be effective.

Management of Functional Safety

The requirements for the Management of Functional Safety are summarised in Table 24.

TABLE 24. REQUIREMENTS FOR THE MANAGEMENT OF FUNCTIONAL SAFETY

Management of Functional Safety Requirement	Description
General Requirements IEC61511-1, 5.2.1 A policy and strategy should be specified together with the means of communication within the organisation.	**Policy and Communications** There should be a Functional Safety Policy in place which should be communicated throughout the organisation. It is recommended that the content of the policy should include specific Functional Safety objectives along with the means of evaluating whether they are achieved and the method of communication within the organisation.
A Functional Safety Management System should be in place to ensure that the SIS has the ability to place and maintain the process in a safe state.	A high level Functional Safety Management document should be available which should identify to all lifecycle phases in scope. The management document should reference the procedures necessary for all safety-related activities. Procedures must be in place to specify all management and technical activities to be carried out on the project. The procedures should identify the documents to be produced. Projects should be controlled using a Quality and Safety Plan which identifies the activities to be carried out, the means of control and which allows sign-off on completion.

Management of Functional Safety Requirement	Description
Organisation and Resources IEC61511-1, 5.2.2 Persons, departments, organisations or other units which are responsible for carrying out and reviewing each of the safety lifecycle phases shall be identified and be informed of the responsibilities assigned to them (including where relevant, licensing authorities or safety regulatory bodies).	**Roles and Responsibilities** All persons, departments and organisations responsible for carrying out and reviewing safety-related activities, should be identified and their responsibilities made clear. Typically, within an organisation, this could be achieved with published organisation charts, identifying individuals and their roles. Job descriptions would then identify responsibilities for each role.
Persons, departments or organisations involved in safety lifecycle activities shall be competent to carry out the activities for which they are accountable.	**Competence** Competence of all responsible persons defined above shall be documented. Procedures should be in place to ensure responsible persons have appropriate competence for the activities assigned to them. The procedure should include a review and assessment of competence and training needs. Documentation of competence should consider: a) engineering knowledge (applicable to the process, the technology, the novelty and complexity of the application, the sensors and final elements); b) adequate management and leadership skill appropriate to the role in the safety lifecycle; c) understanding of the potential consequence of an event; the safety integrity of the SIFs; safety engineering and legal and safety regulatory requirements.
Risk Evaluation and Risk Management IEC61511-1, 5.2.3 Hazards should be identified, risks evaluated and the necessary risk reduction determined.	**SIL Determination** Refer to Section [6].

Management of Functional Safety Requirement	Description
Planning IEC61511-1, 5.2.4 Safety planning shall take place to define the activities that are required to be carried out along with the persons, department, organisation or other units responsible to carry out these activities. This planning shall be updated as necessary throughout the entire safety lifecycle.	**Planning** The planning should ensure that the management of FS, the verification and FS assessment activities are scheduled and applied to the relevant lifecycle phases. Planning may be included in the project quality plan and should identify all safety-related activities, timing and responsible individuals or organisations. Each safety-related activity may include references to procedures or working practices, development or production tools.
Implementing and Monitoring IEC61511-1, 5.2.5 Procedures should be implemented to ensure prompt follow-up and satisfactory resolution of recommendations arising from: a) hazard analysis and risk assessment; b) assessment and audits; c) verification and validation; d) post-incident activities.	**Implementing and Monitoring** The procedures should allow for raising recommendations arising from analysis and review activities and a method for review and tracking recommendations to their resolution should be implemented. There must be a procedure to ensure that any recommendations arising from incidents or hazards, can be acted upon.
Procedures should be implemented to evaluate the performance of the SIS against the safety requirements including: a) Collection and analysis of field failure data during operation; b) Recording of demands on the SIF to ensure that assumptions made during the SIL Determination remain valid.	Where the organization is responsible for the operation and maintenance phases, procedures must be in place for recognising operations and maintenance performance including: • systematic faults; • recurring faults; • assessing demand rates and failure rates in accordance with assumptions during design or FS Assessment. Requirements for FS Audits should include: frequency, independence, required documentation and follow up.

Management of Functional Safety Requirement	Description
Any supplier, providing products or services to an organisation, having overall responsibility for one or more phases of the safety lifecycle, shall deliver products or services as specified by that organisation and shall have an appropriate quality management system. Procedures shall be in place to establish the adequacy of the quality management system.	**Supplier Management** Suppliers shall deliver products as specified and shall have an appropriate QMS. Typically, procurement will be from an approved suppliers list and controlled by procurement specification. Procedures should be in place to audit approval of suppliers.
Assessment, Auditing and Revisions IEC61511-1, 5.2.6 A procedure shall be defined and executed for a functional safety assessment such that a judgement can be made as to the functional safety and safety integrity achieved by the safety instrumented system. The procedure shall require that an assessment team is appointed which includes the technical, application and operations expertise needed for the particular application. The membership of the assessment team shall include at least one senior, competent person not involved in the project design team. The stages in the safety lifecycle at which the functional safety assessment activities are to be carried out shall be identified during safety planning.	**Functional Safety Assessment** FS Assessment activities – refer to section [13]. A procedure should be implemented to enable a FS Assessment to be carried out. The requirements for demonstrating compliance to the SIL and PFD (or PFH) targets established during the SIL Determination [6] are detailed in [13]. A team within the organisation may be appointed provided the requirements for competence and independence are met. Where an external organisation is to be used then the requirements for competence should form part of the supplier management procedure. Requirements for MTR should be included in the scope [8].

Management of Functional Safety Requirement	Description
At least one functional safety assessment should be carried out prior to the identified hazards being present and should confirm: • the hazard and risk assessment have been carried out; • recommendations arising from the hazard and risk assessment have been resolved; • the SIS has been designed, constructed and installed in accordance with the SRS; • the safety, operation and maintenance procedures are in place; • validation activities have been completed; • O&M Training has been completed and appropriate information about the SIS has been provided; • strategies for further assessments are in place.	The FS Assessment should follow a plan for compliance. The points in the project schedule, or safety lifecycle when it should take place should be specified in the project quality and safety plan. It is important that at least one FS Assessment should be carried out prior to the identified hazards being present on the plant or process.
Procedures shall be defined and executed for auditing compliance with requirements including: a) the frequency of the auditing activities; b) the degree of independence between the persons, departments, organisations or other units carrying out the work and those carrying out the auditing activities; c) the recording and follow up activities.	Functional safety audits should be carried out to verify that appropriate procedures are in place on the project and that they have been implemented. Typically, a functional safety audit should be carried out very early in the project lifecycle to ensure that there are procedures in place to cover all of the safety-related activities. Subsequent audits should take place at intervals throughout the project to ensure that the procedures are being followed and that any recommendations or follow-up activities are carried out.

Management of Functional Safety Requirement	Description
SIS Configuration Management IEC61511-1, 5.2.7 Procedures for configuration management of the SIS during the lifecycle, should be available. The following should be specified: a) the stage at which formal configuration control is implemented; b) the method of identification of parts (hardware and software); c) procedures for preventing unauthorized parts from entering service.	**Configuration Management** Procedures for configuration management, initiation of modification, approvals procedure and ensuring follow-up of change requests will probably already exist under a typical QMS. However, when considering changes to a safety function, there must be some form of impact analysis to determine whether the case for safety could be compromised and which point in the lifecycle to go back to, in order to begin the reassessment process. A procedure to conduct the impact analysis and manage the reassessment may be necessary.

Most of the requirements may already be covered by an organisation's Quality Management System (QMS). The following sections highlight some areas that typically need to be addressed.

General Requirements

There must be a policy and strategy for achieving functional safety within the organisation and the means whereby this is communicated throughout the organisation, must be identified.

It is important that the organisation develops its own functional safety policy as this will require stake-holders within the organisation to think carefully about what functional safety means to the organisation, how this can be communicated to create a functional safety culture that reaches throughout the organisation, in all its activities.

Organisation and Resources

All project personnel must be identified based on their competence, and their responsibilities defined. Competence of staff must be recorded in a competence register and there must be a procedure to review competence, to periodically update the register based on experience gained, and to review training needs. Competence requirements must be defined for each project role.

Most organisations new to functional safety may find it beneficial to appoint a Safety Authority (SA) who will have responsibility for functional safety, corporate policy and communications, the lifecycle phases and planning of activities. The SA will be independent of projects.

In all probability, they may also have to establish and manage a competence register or develop an existing system to include functional safety activities and responsibilities.

Project Implementation and Monitoring

If there are some activities that are new to scope, e.g. HAZOP, then a procedure for conducting HAZOPs must be created. If, for example, a development is to include safety-related application software then a procedure for ensuring the software is developed in accordance with Lifecycle Phase 4 [1], must be in place.

Configuration Management and Modification

Procedures for configuration management, initiation of modification, approvals procedure and ensuring follow-up of change requests usually already exist under a typical QMS.

However, when considering changes to a safety function, there must be some form of impact analysis to determine whether the case for safety could be compromised and which point in the lifecycle to go back to, in order to begin the reassessment process. A procedure to conduct the impact analysis and manage the reassessment may be necessary.

O&M Performance

Depending upon the phases of the lifecycle in scope, it may be necessary to implement procedures to deal with, and collect and maintain information arising from: hazards, incidents and modifications. The procedures may also describe:

- handling hazardous incidents;
- analysis of detected hazards;
- verification activities.

Collecting data and maintaining records may be necessary because during the safety assessment, it may have been assumed that the safety function was for example, a demand mode system. Monitoring the demand rate placed on the safety function therefore ensures that the appropriate targets and performance measures were set and remain valid.

20. REFERENCES

20.1. IEC61508:2010, Functional Safety of Electrical/ Electronic/ Programmable Electronic Safety Related Systems.

20.2. IEC615112004: Functional Safety: Safety Instrumented Systems for the Process Industry.

20.3. Health and Safety at Work etc. Act 1974.

20.4. ANSI/ISA-84.00.01-2004 Part 1 (IEC 61511-1 Mod). Functional Safety: Safety Instrumented Systems for the Process Industry Sector.

20.5. Reducing Risks, Protecting People, HSE 2001, ISBN 0 7176 2151 0.

20.6. AIChE Centre for Chemical Process Safety, Layer of Protection Analysis (LOPA), 2001

20.7. IEC61784-3:2010 Industrial Communications Networks. Profiles Part-3: Functional safety Fieldbuses – General Rules and profile Definitions.

20.8. Derivation of the Simplified PFDavg Equations, D Chauhan, Rockwell Automation (FSC).

20.9. General Reliability Calculations for MooN Configurations, KJ Kirkcaldy, Rockwell Automation (FSC).

21. DEFINITIONS

DEFINITIONS	
2oo3	Two out of three logic circuit (2/3 logic circuit) A logic circuit with three independent inputs. The output of the logic circuit is the same state as any two matching input states. For example a safety circuit where three sensors are present and a signal from any two of those sensors is required to call for a shut down. This 2oo3 system is said to be single fault tolerant (HFT = 1) in that one of the sensors can fail dangerously and the system can still safely shut down. Other voting systems include 1oo1, 1oo2, 2oo2, 1oo3 and 2oo4.
IEC61508	The IEC standard covering Functional Safety of electrical / electronic / programmable electronic safety-related systems The main objective of IEC61508 is to use safety instrumented systems reduce risk to a tolerable level by following the overall, hardware and software safety lifecycle procedures and by maintaining the associated documentation. Issued in 1998 and 2000, it has since come to be used mainly by safety equipment suppliers to show that their equipment is suitable for use in safety integrity level rated systems.
IEC61511	The IEC standard for use of electrical / electronic / programmable electronic safety-related systems in the process industry. Like IEC 61508 it focuses on a set of safety lifecycle processes to manage process risk. It was originally published by the IEC in 2003 and taken up by the US in 2004 as ISA 84.00.01- 2004. Unlike IEC 61508, this standard is targeted toward the process industry users of safety instrumented systems.
ALARP	As low as reasonably practicable. The philosophy of dealing with risks that fall between an upper and lower extreme. The upper extreme is where the risk is so great that it is rejected completely while the lower extreme is where the risk is, or has been made to be, insignificant. This philosophy considers both the costs and benefits of risk reduction to make the risk "as low as reasonably practicable".
Architectural constraints	Limitations that are imposed on the hardware selected to implement a safety-instrumented function, regardless of the performance calculated for a subsystem. Architectural constraints are specified (in IEC61508-2-Table 2 and IEC 61511-Table 5) according to the required SIL of the subsystem, type of components used, and SFF of the subsystem's components. Type A components are simple devices not incorporating microprocessors, and Type B devices are complex devices such as those incorporating microprocessors. See Fault Tolerance.
Architecture	The voting structure of different elements in a safety instrumented function. See Architectural Constraints, Fault Tolerance and 2oo3.

DEFINITIONS	
Availability	The probability that a device is operating successfully at a given moment in time. This is a measure of the "uptime" and is defined in units of percent. For most tested and repaired safety system components, the availability varies as a saw tooth with time as governed by the proof test and repair cycles. Thus the integrated average availability is used to calculate the average probability of failure on demand. See PFDavg.
Basic process control system	System which responds to input signals from the process, associated equipment, and/or an operator and generates output signals causing the process and its associated equipment to operate in the desired way. The BPCS cannot perform any safety instrumented functions rated with a safety integrity level of 1 or better unless it meets proven in use requirements. See proven in use.
BPCS	See Basic Process Control System.
Cause and effect diagram	One method commonly used to show the relationship between the sensor inputs to a safety function and the required outputs. Often used as part of a safety requirements specification. The method's strengths are a low level of effort and clear visual representation while its weaknesses are a rigid format (some functions cannot be represented w/ C-E diagrams) and the fact that it can oversimplify the function.
Common mode failure	A random stress that causes two or more components to fail at the same time for the same reason. It is different from a systematic failure in that it is random and probabilistic but does not proceed in a fixed, predictable, cause and effect fashion. See systematic failure.
Consequence	The magnitude of harm or measure of the resulting outcome of a harmful event. One of the two components used to define a risk.
Proof test coverage	The percentage failures that are detected during the servicing of equipment. In general it is assumed that when a proof test is performed any errors in the system are detected and corrected (100% proof test coverage).
D Diagnostics	Some safety rated logic solvers are designated as having capital D diagnostics. These are different from regular diagnostics in that the unit is able to reconfigure its architecture after a diagnostic has detected a failure. The greatest effect is for 1oo2D systems which can reconfigure to 1oo1 operation upon detecting a safe failure. Thus the spurious trip rate for such a system is dramatically reduced.
Dangerous failure	A failure of a component in a safety instrumented function that prevents that function from achieving a safe state when it is required to do so. See failure mode.
Diagnostic coverage	A measure of a system's ability to detect failures. This is a ratio between the failure rates for detected failures to the failure rate for all failures in the system.

DEFINITIONS	
E/E/PE Electrical / Electronic / Programmable Electronic	See 61508 and 61511.
Event tree analysis	A method of fault propagation modelling. The analysis constructs a tree-shaped picture of the chains of events leading from an initiating event to various potential outcomes. The tree expands from the initiating event in branches of intermediate propagating events. Each branch represents a situation where a different outcome is possible. After including all of the appropriate branches, the event tree ends with multiple possible outcomes.
Fail close	A condition wherein the valve closing component moves to a closed position when the actuating energy source fails.
Fail open	A condition wherein the valve closing component moves to an open position when the actuating energy source fails.
Fail safe (or preferably de-energize to trip)	A characteristic of a particular device which causes that device to move to a safe state when it loses electrical or pneumatic energy.
Failure modes	The way that a device fails. These ways are generally grouped into one of four failure modes: Safe Detected (SD), Dangerous Detected (DD), Safe Undetected (SU), and Dangerous Undetected (DU) per ISA TR84.0.02.
Failure rate	The number of failures per unit time for a piece of equipment. Usually assumed to be a constant value. It can be broken down into several categories such as safe and dangerous, detected and undetected, and independent/normal and common cause. Care must be taken to ensure that burn in and wear-out are properly addressed so that the constant failure rate assumption is valid.
Fault tolerance	Ability of a functional unit to continue to perform a required function in the presence of random faults or errors. For example a 1oo2 voting system can tolerate one random component failure and still perform its function. Fault tolerance is one of the specific requirements for safety integrity level (SIL) and is described in more detail in IEC61508 Part 2 Tables 2 and 3 and in IEC61511 (ISA 84.01 2004) in Clause 11.4
Fault tree diagram	Probability combination method for estimating complex probabilities. Since it generally takes the failure view of a system, it is useful in multiple failure mode modelling. Care must be taken when using it to calculate integrated average probabilities.
FMECA	Failure Modes Effects and Criticality Analysis - This is a detailed analysis of the different failure modes and criticality analysis for a piece of equipment.
Functional safety	Freedom from unacceptable risk achieved through the safety lifecycle. See IEC61508, IEC65111, safety lifecycle, and tolerable risk.
Hazard	The potential for harm.

DEFINITIONS	
HAZOP	Hazards and operability study. A process hazards analysis procedure originally developed by ICI in the 1970s. The method is highly structured and divides the process into different operationally-based nodes and investigates the behaviour of the different parts of each node based on an array of possible deviation conditions or guidewords.
HFT	Hardware fault tolerance (see fault tolerance)
HSE (UK)	Health and Safety Executive
IEC	International Electrotechnical Commission. A worldwide organization for standardization. The object of the IEC is to promote international cooperation on all questions concerning standardization in the electrical and electronic fields. To this end and in addition to other activities, the IEC publishes international standards. See 61508 and 61511. Impact analysis activity of determining the effect that a change to a function or component will have to other functions or components in that system as well as to other systems
Incident	The result of an initiating event that is not stopped from propagating. The incident is most basic description of an unwanted accident, and provides the least information. The term incident is simply used to convey the fact that the process has lost containment of the chemical or other potential energy source. Thus the potential for causing damage has been released but its harmful result has not has not taken specific form.
IPL	Independent protection layer or layers. This refers to various other methods of risk reduction possible for a process. Examples include items such as rupture disks and relief valves which will independently reduce the likelihood of the hazard escalating into a full accident with a harmful outcome. In order to be effective, each layer must specifically prevent the hazard in question from causing harm, act independently of other layers, have a reasonable probability of working, and be able to be audited once the plant is operation relative to its original expected performance.
Lambda	Failure rate for a system. See failure rate.
Likelihood	The frequency of a harmful event often expressed in events per year or events per million hours. One of the two components used to define a risk. Note that this is different from the traditional English definition that means probability.
LOPA	Layer of Protection Analysis. A method of analysing the likelihood (frequency) of a harmful outcome event based on an initiating event frequency and on the probability of failure of a series of independent layers of protection capable of preventing the harmful outcome.

DEFINITIONS	
Mode (Continuous)	When demands to activate a safety function (SIF) are frequent compared to the test interval of the SIF. Note that other sectors define a separate high demand mode, based on whether diagnostics can reduce the accident rate. In either case, the continuous mode is where the frequency of an unwanted accident is essentially determined by the frequency of a dangerous SIF failure. When the SIF fails, the demand for its action will occur in a much shorter time frame than the function test, so speaking of its failure probability is not meaningful. Essentially all of the dangerous faults of a SIF in continuous mode service will be revealed by a process demand instead of a function test. See low demand mode, high demand mode, and SIL.
Mode (High Demand)	(also continuous mode per IEC61511) Similar to continuous mode only there is specific credit taken for automatic diagnostics. The split between high demand and continuous mode is whether the automatic diagnostics are run many times faster than the demand rate on the safety function. If the diagnostics are slower than this there is no credit for them and the continuous mode applies.
Mode (Low Demand)	(also demand mode per IEC61511) when demands to activate the safety instrumented function (SIF) are infrequent compared to the test interval of the SIF. The process industry defines this mode when the demands to activate the SIF are less than once every two proof test intervals. The low demand mode of operation is the most common mode in the process industries. When defining safety integrity level for the low demand mode, a SIF's performance is measured in terms of average Probability of Failure on Demand (PFDavg). In this demand mode, the frequency of the initiating event, modified by the SIF's probability of failure on demand times the demand rate and any other downstream layers of protection determine the frequency of unwanted accidents.
MTTR	Mean Time to Repair – The average time between the occurrence of a failure and the completion of the repair of that failure. This includes the time needed to detect the failure, initiate the repair and fully complete the repair.
Occupancy	A measure of the probability that the effect zone of an accident will contain one or more personnel receptors of the effect. This probability should be determined using plant-specific staffing philosophy and practice.
P&ID	Piping and instrumentation drawing. Shows the interconnection of process equipment and the instrumentation used to control the process. In the process industry, a standard set of symbols is used to prepare drawings of processes. The instrument symbols used in these drawings are generally based on Instrument Society of America (ISA) Standard S5. 1. 2. The primary schematic drawing used for laying out a process control installation.

DEFINITIONS	
PFDavg	Probability of Failure on Demand average- This is the probability that a system will fail dangerously, and not be able to perform its safety function when required. PFD can be determined as an average probability or maximum probability over a time period. IEC61508/61511 and ISA 84.01 use PFDavg as the system metric upon which the SIL is defined.
Proof test	Testing of safety system components to detect any failures not detected by automatic on-line diagnostics i.e. dangerous failures, diagnostic failures, parametric failures followed by repair of those failures to an equivalent as new state. Proof testing is a vital part of the safety lifecycle and is critical to ensuring that a system achieves its required safety integrity level throughout the safety lifecycle.
Protection layer	See IPL.
Proven in use	Basis for use of a component or system as part of a safety integrity level (SIL) rated safety instrumented system (SIS) that has not been designed in accordance with IEC61508. It requires sufficient product operational hours, revision history, fault reporting systems, and field failure data to determine if the is evidence of systematic design faults in a product. IEC 61508 provides levels of operational history required for each SIL.
Proof Test Interval	The time interval between servicing of the equipment.
Random failure	A failure occurring at a random time, which results from one or more degradation mechanisms. Random failures can be effectively predicted with statistics and are the basis for the probability of failure on demand based calculations requirements for safety integrity level. See systematic failure.
Redundancy	Use of multiple elements or systems to perform the same function. Redundancy can be implemented by identical elements (identical redundancy) or by diverse elements (diverse redundancy). Redundancy of primarily used to improve reliability or availability.
Reliability	1. The probability that a device will perform its objective adequately, for the period of time specified, under the operating conditions specified. 2. The probability that a component, piece of equipment or system will perform its intended function for a specified period of time, usually operating hours, without requiring corrective maintenance.
Reliability block diagram	Probability combination method for estimating complex probabilities. Since it generally takes the "success" view of a system, it can be confusing when used in multiple failure mode modelling.
RRF	Risk Reduction Factor -The inverse of PFDavg

DEFINITIONS	
Safe failure	Failure that does not have the potential to put the safety instrumented system in a dangerous or fail-to-function state. The situation when a safety related system or component fails to perform properly in such a way that it calls for the system to be shut down or the safety instrumented function to activate when there is no hazard present.
Safe failure fraction	See SFF.
Safe state	The state of the process after acting to remove the hazard resulting in no significant harm.
SFF	Safe Failure Fraction - The fraction of the overall failure rate of a device that results in either a safe fault or a diagnosed (detected) unsafe fault. The safe failure fraction includes the detectable dangerous failures when those failures are annunciated and procedures for repair or shutdown are in place.
SIF	Safety Instrumented Function – A set of equipment intended to reduce the risk due to a specific hazard (a safety loop). Its purpose is to: 1. Automatically taking an industrial process to a safe state when specified conditions are violated; 2. Permit a process to move forward in a safe manner when specified conditions allow (permissive functions); or 3. Taking action to mitigate the consequences of an industrial hazard. It includes elements that detect an accident is imminent, decide to take action, and then carry out the action needed to bring the process to a safe state. Its ability to detect, decide and act is designated by the safety integrity level (SIL) of the function. See SIL.
SIL	Safety Integrity Level - A quantitative target for measuring the level of performance needed for safety function to achieve a tolerable risk for a process hazard. Defining a target SIL level for the process should be based on the assessment of the likelihood that an incident will occur and the consequences of the incident. The following table describes SIL for different modes of operation.
SIL verification	The process of calculating the average probability of failure on demand (or the probability of failure per hour) and architectural constraints for a safety function design to see if it meets the required SIL.
SIS	Safety Instrumented System – Implementation of one or more Safety Instrumented Functions. A SIS is composed of any combination of sensor(s), logic solver(s), and final element(s). A SIS is usually has a number of safety functions with different safety integrity levels (SIL) so it is best avoid describing it by a single SIL. See SIF.
Spurious trip	See Safe failure

DEFINITIONS	
Systematic failure	A failure that happens in a deterministic (non random) predictable fashion from a certain cause, which can only be eliminated by a modification of the design or of the manufacturing process, operational procedures, documentation, or other relevant factors. Since these are not mathematically predictable, the safety lifecycle includes a large number of procedures to prevent them from occurring. The procedures are more rigorous for higher safety integrity level systems and components. Such failures cannot be prevented with simple redundancy.

22. INDEX